Set-Valued Analysis

Set-Valued Analysis

Editors

Anca Croitoru
Anna Rita Sambucini
Bianca Satco

MDPI • Basel • Beijing • Wuhan • Barcelona • Belgrade • Manchester • Tokyo • Cluj • Tianjin

Editors
Anca Croitoru
University "Alexandru Ioan Cuza" of Iasi
Romania

Anna Rita Sambucini
University of Perugia 1
Italy

Bianca Satco
University of Suceava
Romania

Editorial Office
MDPI
St. Alban-Anlage 66
4052 Basel, Switzerland

This is a reprint of articles from the Special Issue published online in the open access journal *Mathematics* (ISSN 2227-7390) (available at: https://www.mdpi.com/journal/mathematics/special_issues/Set_Valued_Analysis).

For citation purposes, cite each article independently as indicated on the article page online and as indicated below:

LastName, A.A.; LastName, B.B.; LastName, C.C. Article Title. *Journal Name* **Year**, *Volume Number*, Page Range.

ISBN 978-3-0365-0784-2 (Hbk)
ISBN 978-3-0365-0785-9 (PDF)

© 2021 by the authors. Articles in this book are Open Access and distributed under the Creative Commons Attribution (CC BY) license, which allows users to download, copy and build upon published articles, as long as the author and publisher are properly credited, which ensures maximum dissemination and a wider impact of our publications.

The book as a whole is distributed by MDPI under the terms and conditions of the Creative Commons license CC BY-NC-ND.

Contents

About the Editors vii

Preface to "Mathematics Set-Valued Analysis" ix

Luisa Di Piazza and Kazimierz Musiał
Decompositions of Weakly Compact Valued Integrable Multifunctions
Reprinted from: *Mathematics* 2020, *8*, 863, doi:10.3390/math8060863 1

Antonio Boccuto, Bipan Hazarika and Hemanta Kalita
Kuelbs–Steadman Spaces for Banach Space-Valued Measures
Reprinted from: *Mathematics* 2020, *8*, 1005, doi:10.3390/math8061005 15

Charles Castaing, Christiane Godet-Thobie and Le Xuan Truong
Fractional Order of Evolution Inclusion Coupled with a Time and State Dependent Maximal
Monotone Operator
Reprinted from: *Mathematics* 2020, *8*, 1395, doi:10.3390/math8091395 - 31

Abbas Ghaffari, Reza Saadati, Radko Mesiar
Inequalities in Triangular Norm-Based $*$-Fuzzy $(L^+)^p$ Spaces
Reprinted from: *Mathematics* 2020, *8*, 1984, doi:10.3390/math8111984 63

Bianca Satco and George Smyrlis
Applications of Stieltjes Derivatives to Periodic Boundary Value Inclusions
Reprinted from: *Mathematics* 2020, *8*, 2142, doi:0.3390/math8122142 85

Diana Caponetti, Mieczysław Cichoń, Valeria Marraffa
On Regulated Solutions of Impulsive Differential Equations with Variable Times
Reprinted from: *Mathematics* 2020, *8*, 2164, doi:10.3390/math8122164 109

Danilo Costarelli, Anca Croitoru, Alina Gavriluţ, Alina Iosif and Anna Rita Sambucini
The Riemann-Lebesgue Integral of Interval-Valued Multifunctions
Reprinted from: *Mathematics* 2020, *8*, 2250, doi:10.3390/math8122250 125

Reny George and Hemanth Kumar Pathak
Some New Extensions of Multivalued Contractions in a b-metric Space and Its Applications
Reprinted from: *Mathematics* 2021, *9*, 12, doi:10.3390/math9010012 143

About the Editors

Anca Croitoru has a Ph.D. in Mathematics (since 2000) from the University Alexandru Ioan Cuza of Iasi, Romania and is currently an Associate Professor at the University Alexandru Ioan Cuza of Iasi. She has published more than 50 papers in journals and conference proceedings and participated in more than 50 international and Romanian conferences.

Anna Rita Sambucini is an Associate Professor at the University of Perugia, Italy. Her research activity focuses mainly measure theory and its applications, with particular attention to finitely additive measures, vector lattices, and multifunctions. She has published more than 60 papers in journals and conference proceedings and she is associate editor for the Australian Journal of Mathematical Analysis and Applications and Open Mathematics.

Bianca Satco obtained a Ph.D. in Mathematics in 2005 at Université de Bretagne Occidentale in Brest, France, in cooperation with the University Alexandru Ioan Cuza in Iasi, Romania. She has led two national research projects, published more than 50 articles in journals indexed in Web of Science, ZentralBlatt fur Mathematik, and Mathematical Reviews databases, and given talks at more than 30 international meetings in Romania and abroad. Concerning her editorial activity, she is a reviewer for ZentralBlatt fur Mathematik and Mathematical Reviews (since 2007) and a referee for more than 20 well-known journals (e.g., Journal of Mathematical Analysis and Applications, Mathematische Nachrichten, Bulletin des Sciences Mathematiques, Applied Mathematics and Computation, Glasgow Mathematical Journal, Set-Valued and Variational Analysis, Mediterranean Journal of Mathematics, Boundary Value Problems, Advances in Difference Equations, Electronic Journal of Differential Equations).

Preface to "Mathematics Set-Valued Analysis"

Set-valued analysis is an important and somehow strange field, with unexpected applications in economics, game theory, decision making, nonlinear programming, biomathematics and statistics.

This book highlights some interesting subjects in set-valued analysis, both theoretical and practical. These topics cover various areas, such as set-valued measures and integrals, applications to differential inclusions and decision making, and related topics in measure theory.

Anca Croitoru, Anna Rita Sambucini, Bianca Satco
Editors

Article

Decompositions of Weakly Compact Valued Integrable Multifunctions

Luisa Di Piazza [1],* and Kazimierz Musiał [2]

1. Department of Mathematics, University of Palermo, Via Archirafi 34, 90123 Palermo, Italy
2. Institut of Mathematics, Wrocław University, Pl. Grunwaldzki 2/4, 50-384 Wrocław, Poland; kazimierz.musial@math.uni.wroc.pl
* Correspondence: luisa.dipiazza@unipa.it

Received: 21 April 2020; Accepted: 21 May 2020; Published: 26 May 2020

Abstract: We give a short overview on the decomposition property for integrable multifunctions, i.e., when an "integrable in a certain sense" multifunction can be represented as a sum of one of its integrable selections and a multifunction integrable in a narrower sense. The decomposition theorems are important tools of the theory of multivalued integration since they allow us to see an integrable multifunction as a translation of a multifunction with better properties. Consequently, they provide better characterization of integrable multifunctions under consideration. There is a large literature on it starting from the seminal paper of the authors in 2006, where the property was proved for Henstock integrable multifunctions taking compact convex values in a separable Banach space X. In this paper, we summarize the earlier results, we prove further results and present tables which show the state of art in this topic.

Keywords: gauge multivalued integral; scalarly defined multivalued integral; decomposition of a multifunction

MSC: 28B20; 26E25; 26A39; 28B05; 46G10; 54C60; 54C65

1. Introduction

Various investigations in mathematical economics, optimal control and multivalued image reconstruction led to study of the integrability of multifunctions. In fact, the multivalued integration has shown to be a useful tool when modeling theories in different fields [1–7]. Also, the study of multivalued integrals arises in a natural way in connection with statistical problems (see, for example, [8–10]). But the topic is interesting also from the point of view of measure and integration theory, as we can see in the papers [1,7,11–38].

Here we examine two groups of the integrals: those functionally determined (we call them "scalarly defined integrals") (as Pettis, Henstock–Kurzweil–Pettis, Denjoy–Pettis integrals) and those identified by gauges (we call them "gauge defined integrals") as Henstock, McShane and Birkhoff integrals. The last class also includes versions of Henstock and McShane integrals, when only measurable gauges are allowed, and the variational Henstock and McShane integrals. We investigate only multifunctions with weakly compact and convex values. More general theory of integration is not sufficiently developed until now.

In particular, decomposition properties are considered both for scalarly defined integrals and for gauge defined integrals. The results presented here are contained in some papers quoted in the bibliography or can be easily obtained. Only some results are discussed. The novelty of the present article relies in the fact that we sumarize the results known until now in the field. Moreover, we compare them and in Table 2A,B we provide a clear view of the state of art in the topic.

2. Preliminaries

Throughout the paper X is a Banach space with norm $\|\cdot\|$ and its dual X^*. The closed unit ball of X is denoted B_X. The symbol $cwk(X)$ denotes the collection of all nonempty convex weakly compact subsets of X. For every $C \in cwk(X)$ the *support function of* C is denoted by $s(\cdot, C)$ and defined on X^* by $s(x^*, C) = \sup\{\langle x^*, x\rangle \colon x \in C\}$, for each $x^* \in X^*$. We set $\|C\|_h = d_H(C, \{0\}) := \sup\{\|x\| : x \in C\}$, where d_H is the Hausdorff metric on the hyperspace $cwk(X)$. Let $([0,1], \lambda, \mathcal{L})$ be the unit interval equipped with Lebesgue measure λ and Lebesgue measurable sets $\mathcal{L}$, while $\mathcal{I}$ is the collection of all closed subintervals of $[0,1]$. L_0 is the collection of all strongly measurable X-valued functions defined on $[0,1]$. Unless otherwise noted, all investigated multifunctions are defined on $[0,1]$ and take values in $cwk(X)$. A function $f : [0,1] \to X$ is called a *selection of* a multifunction Γ if $f(t) \in \Gamma(t)$, for almost every $t \in [0,1]$.

We recall that if $\Phi : \mathcal{L} \to Y$ is an additive vector measure with values in a normed space Y, then the *variation of* Φ is the extended non negative function $|\Phi|$ whose value on a set $E \in \mathcal{L}$ is given by $|\Phi|(E) = \sup_\pi \sum_{A \in \pi} \|\Phi(A)\|$, where the supremum is taken over all partitions π of E into a finite number of pairwise disjoint members of $\mathcal{L}$. If $|\Phi| < \infty$, then Φ is called a *measure of finite variation*. If Φ is defined only on $\mathcal{I}$, the finite partitions considered in the definition of variation are composed by intervals. In this case we will speak of *finite interval variation* and we will use the symbol $\widetilde{\Phi}$, namely:

$$\widetilde{\Phi}([0,1]) = \sup\{\sum_i \|\Phi(I_i)\| \colon \{I_1, \ldots, I_n\} \text{ is a finite interval partition of } [0,1]\}.$$

If $\{I_1, \ldots, I_p\}$ is a partition in $[0,1]$ into intervals and $t_j \in [0,1]$, $j = 1, \ldots, p$, then $\{(I_j, t_j)\}_{j=1}^p$ is called an $\mathcal{I}$-partition. If δ is a gauge (that is positive function) on $[0,1]$ and $I_j \subset [t_j - \delta(t_j), t_j + \delta(t_j)], j = 1, \ldots, p$, $p \in \mathbb{N}$, then the $\mathcal{I}$-partition is called δ-fine.

Moreover a usefull tool in our investigation is the notion of variational measure generated by an interval multimeasure. Given an interval multimeasure $\Phi : \mathcal{I} \to cwk(X)$, we call *variational measure* $V_\Phi : \mathcal{L} \to \mathbb{R}$ generated by Φ, the measure whose value on a set $E \in \mathcal{L}$ is given by

$$V_\Phi(E) := \inf_\delta \{Var(\Phi, \delta, E) : \delta \text{ is a gauge on } E\},$$

where

$$Var(\Phi, \delta, E) = \sup\left\{\sum_{j=1}^p \|\Phi(I_j)\|_h \colon \{(I_j, t_j)\}_{j=1}^p \text{ is a } \delta\text{-fine } \mathcal{I}\text{-partition, with } t_j \in I_j \cap E, j = 1, \ldots, p\right\}.$$

Now we recall here briefly the definitions of the integrals involved in this article. A scalarly integrable multifunction $\Gamma : [0,1] \to cwk(X)$ is *Pettis integrable* (P_e) in $cwk(X)$, if for every set $A \in \mathcal{L}$ there exists a set $M_\Gamma(A) \in cwk(X)$ such that $s(x^*, M_\Gamma(A)) = \int_A s(x^*, \Gamma) \, d\lambda$ for every $x^* \in X^*$. We write it as $(P)\int_A \Gamma \, d\lambda$ or $M_\Gamma(A)$. A multifunction $\Gamma : [0,1] \to cwk(X)$ is called *Bochner integrable* if it is Bochner measurable (i.e., there exists a sequence of simple multifunctions $\Gamma_n : [0,1] \to cwk(X)$ such that for almost all $t \in [0,1]$ one has $\lim_n d_H(\Gamma_n(t), \Gamma(t)) = 0$) and integrably bounded. We will denote the family by L_1.

A multifunction $\Gamma : [0,1] \to cwk(X)$ is said to be *McShane* (MS) (resp. *Henstock* (H)) integrable on $[0,1]$, if there exists $\Phi_\Gamma([0,1]) \in cwk(X)$ with the property that for every $\varepsilon > 0$ there exists a gauge δ on $[0,1]$ such that for each δ-fine $\mathcal{I}$-partition $\{(I_1, t_1), \ldots, (I_p, t_p)\}$ of $[0,1]$ (with $t_i \in I_i$ for all i), we have

$$d_H\left(\Phi_\Gamma([0,1]), \sum_{i=1}^p \Gamma(t_i)\lambda(I_i)\right) < \varepsilon. \qquad (1)$$

If the gauges above are taken to be measurable, then we speak of $\mathcal{H}$ (resp. *Birkhoff*)-integrability on $[0,1]$. If $I \in \mathcal{I}$, then $\Phi_\Gamma(I) := \Phi_{\Gamma\chi_I}[0,1]$.

Finally if, instead of Formula (1), we have

$$\sum_{i=1}^{p} d_H\left(\Phi_\Gamma(I_i), \Gamma(t_i)\lambda(I_i)\right) < \varepsilon, \qquad (2)$$

we speak about *variational Henstock (vH)* (resp. *McShane (vMS)*) integrability on $[0,1]$.

The definition of variational Henstock (resp. McShane) integral comes from the classical Saks-Henstock Lemma for real valued functions. In case of Banach valued functions, they coincide with the definitions of Henstock (resp. McShane) integral if and only if the Banach space is of finite dimension. In the other cases, the variational integrals possesse better properties than Henstock or McShane integrals. In particular, the notion of variational Henstock integrability is a usefull tool to study the diferrentiability of Pettis integrals (cf. [13] (Corollary 4.1)). Formula (2) is the natural extension of such integrals to the multivalued case.

Moreover by [18] (Theorem 6.6) vH-integrability and $v\mathcal{H}$ integrability coincide. In all the cases $\Phi_\Gamma : \mathcal{I} \to cwk(X)$ is an additive interval multimeasure. A multifunction $\Gamma : [0,1] \to cwk(X)$ is said to be *Henstock-Kurzweil-Pettis (HKP)* integrable in $cwk(X)$ if it is scalarly Henstock-Kurzweil (HK)-integrable and for each $I \in \mathcal{I}$ there exists a set $N_\Gamma(I) \in cwk(X)$ such that $s(x^*, N_\Gamma(I)) = (HK)\int_I s(x^*, \Gamma(t))dt$ for every $x^* \in X^*$. If an HKP-integrable Γ is scalarly integrable, then it is called *weakly McShane integrable (wMS)*. We recall that a function $f : [0,1] \to \mathbb{R}$ is Denjoy-Khintchine (DK) integrable ([39] (Definition 11)), if there exists an ACG function (cf. [39]) F such that its approximate derivative is almost everywhere equal to f.

A multifunction $\Gamma : [0,1] \to cwk(X)$ is *Denjoy-Khintchine-Pettis (DKP)* integrable in $cwk(X)$, if for each $x^* \in X^*$ the function $s(x^*, \Gamma(\cdot))$ is Denjoy-Khintchine integrable and for every $I \in \mathcal{I}$ there exists $C_I \in cwk(X)$ with $(DK)\int_I s(x^*, \Gamma(t))dt = s(x^*, C_I)$, for every $x^* \in X^*$.

A multifunction $\Gamma : [0,1] \to cwk(X)$ satisfies the *Db-condition* (resp. *DL-condition*) if

$$\sup\mathrm{ess}_t\, \mathrm{diam}(\Gamma(t)) < \infty \quad (\text{resp. } \overline{\int}_0^1 \mathrm{diam}(\Gamma(t))dt < +\infty, \text{ where } \overline{\int} \text{ denotes the upper integral}).$$

We say that a multifunction $\Gamma : [0,1] \to cwk(X)$ is *positive* if $s(x^*, \Gamma(\cdot)) \geq 0$ a.e. for each $x^* \in X^*$ separately. Of course, if $0 \in \Gamma(t)$ for almost every $t \in [0,1]$, then Γ is positive. As regards other definitions of measurability and integrability that are treated here and are not explained and the known relations among them, we refer to [3,15–20,26,36,38,40–42], in order not to burden the presentation.

3. Intersections

In this section we are going to highlight some relations among gauge integrability and functionally defined integrability for multifunctions in order to understand better the examples given before. Since we have inclusions

$$DP \supset HKP \supset wMS \supset Pe \quad \text{and} \quad DP \supset HKP \supset H \supset \mathcal{H} \supset vH = v\mathcal{H}$$

only the pairs of different types of integrals are interesting. For what concernes the symbol subscript fv it means that the corresponding integral is of finite variation.

In Table 1 Henstock, $\mathcal{H}$ and $v\mathcal{H}$-integrable functions possessing integrals of finite variation, are not taken into consideration. The reason is simple. In [21] (Theorem 4.5) it is proven that such multifunctions are McShane and Birkhoff integrable, respectively. For a similar reason wMS-integrable multifunctions with integrals of finite variation are omitted. Φ is the indefinite integral of G.

Table 1. Intersections. Arbitrary G.

$\mathcal{GG}$	vH	$\mathcal{H}$	H
P_{efv}	L_1 Remark 1	Bi_{fv} [18] (Theorem 4.3)	MS_{fv} [18] (Theorem 3.3)
P_e	$P_e \cap L_0 + V_\Phi \ll \lambda$ Remark 1	Bi [18] (Theorem 4.3)	MS [18] (Theorem 3.3)
wMS	$P_e \cap L_0 + V_\Phi \ll \lambda$ if $c_0 \not\subseteq X$ Remark 1	Bi if $c_0 \not\subseteq X$ Remark 1	MS if $c_0 \not\subseteq X$ Remark 1

Remark 1. *Observe that, using the Rådström embedding: $i : cwk(X) \to l_\infty(B_{X^*})$ (see for example [43] or [19]) given by $i(A) := s(\cdot, A)$, we have that:*

1. *directly from the definitions and the Rådström embedding, a multifunction $G : [0,1] \to cwk(X)$ is Birkhoff (resp. Henstock, McShane, variationally Henstock) integrable if and only if $i \circ G$ is integrable in the same sense. For the Pettis integrability this is not true. However, for Bochner measurable multifunctions, we have that since $\{G(E) : E \in \mathcal{L}\}$ is separable for the Hausdorff distance and then G is Pettis integrable if and only if $i \circ G$ is Pettis integrable ([26] (Proposition 4.5)), so we have $P_e = MS = Bi$ (for strongly measurable vector valued functions, Pettis, McShane and Birkhoff integrability coincide (see [44] (Corollary 4C) and [45] (Theorem 10)).*

2. *$P_{efv} \cap vH = L_1$ in Table 1 solves the problem of [46], where the authors noticed that $P_e \cap vH \neq L_1$ in case of functions. The inclusion $P_{efv} \cap vH \supset L_1$ is clear. To prove the inclusion $P_{efv} \cap vH \subset L_1$ take $G \in P_{efv} \cap vH$. Then $i \circ G$ is strongly measurable ([17] (Proposition 2.8)) and vH-integrable. If M_G is the Pettis integral of G, then $i \circ M_G$ is a measure of finite variation and $i \circ M_G(I) = (vH) \int_I i \circ G$. It follows that $i \circ G$ is Pettis integrable and then Bochner integrable by [47] (Theorem 4.1) or [48] (Lemma 2). Now we may apply [17] (Proposition 3.6) to obtain variational McShane integrability of G.*

3. *The results for the P_e row and vH column follow from Remark 1, by [17] (Theorem 4.3, d) $\Leftrightarrow$ e)) and [13] (Corollary 4.1), since G is vH integrable if and only if the variational measure V_Φ of its multivalued Pettis integral Φ is λ-continuous ([19] (Theorem 3.3)). Example 1 shows what can happen in the $P_e \setminus vH$ case.*

4. *The results given in wMS row follow from the P_e row and [49] (Theorem 18) or [50] (Theorem 4.4).*

Example 1. *There exists a Pettis integrable multifunction $G : [0,1] \to cwk(X)$ such that $0 \in G(t)$ for every $t \in [0,1]$ and the variational measure associated to its Pettis integral $V_{M_G} \not\ll \lambda$.*

Proof. Let $g : [0,1] \to X$ be a Pettis integrable function such that the variational measure associated to its Pettis integral $V_{\nu_g} \not\ll \lambda$, where $\nu_g(E) = (P) \int_E g \, d\lambda$, (for the existence see [13] (Corollary 4.2, Remark 4.3)), then we take $G(t) := conv\{0, g(t)\}$. The multifunction G is Pettis integrable and $V_{\nu_g}(E) \leq V_{M_G}(E)$ ([17] (Proposition 2.7)). It follows that $V_{M_G} \not\ll \lambda$. □

4. Decompositions

The decomposition of a multifunction Γ integrable in a certain sense into a sum of one of its integrable selections and a multifunction integrable in a narrower sense, relies essentially in the two facts:

(1) Existence of a selection of Γ integrable in the same sense as Γ.
(2) A particular behaviour with respect to the integration of a positive multifunction.

In particular, regarding the results on the existence of selections we can observe that:

Proposition 1. *Let X be any Banach space and let $\Gamma : [0,1] \to cwk(X)$.*

(i) If Γ is Pettis (resp. HKP, wMS or DKP) integrable in $cwk(X)$, then each scalarly measurable selection of Γ is Pettis (resp. HKP, wMS or DK) integrable (see [26] (Corollary 2.3, Theorem 2.5) and [31] (Proposition 3, Remark 3)));

(ii) if Γ is Henstock (resp. McShane) integrable, then it possesses at least one Henstock (resp. McShane) integrable selection (see [33] (Theorem 3.1) or [30] (Theorem 2) in case of a separable X and compact valued Γ);
(iii) if Γ is $\mathcal{H}$ (resp. Birkhoff) integrable, then it possesses at least one $\mathcal{H}$ (resp. Birkhoff) integrable selection (see [17,30] (Theorem 3.4), [18] (Proposition 4.1));
(iv) if Γ is vH integrable, then there exists at least one vH integrable selection (see [18] (Theorem 5.1)); if Γ takes convex compact values and is Bochner integrable, then it possesses at least one Bochner integrable selection (see [17] (Theorem 3.9)).

While, for positive multifunctions, the following relations are known:

Proposition 2. *Let X be any Banach space and let $G : [0,1] \to cwk(X)$. Then*

(i) *If G is Henstock integrable (resp. $\mathcal{H}$-integrable) and positive, then it is also McShane (resp. Birkhoff) integrable on $[0,1]$ (see [18] (Proposition 3.1));*
(ii) *If G is variationally Henstock integrable and positive, then G is Birkhoff integrable (see [17] (Proposition 4.1));*
(iii) *If G is HKP (resp. DKP) integrable and positive, then G is Pettis integrable (see [31] (Lemma 1)).*

In general it is not possible to write $\Gamma = G + f$ with the meaning explained before. We present below a few examples.

Example 2. *There exists a Pettis integrable multifunction $G : [0,1] \to cwk(X)$ such that $0 \in G(t)$ for every $t \in [0,1]$, but G is not McShane integrable.*

Proof. Let $g : [0,1] \to X$ be Pettis but not McShane integrable and let $G(t) := conv\{0, g(t)\}$ be the multifunction determined by g. Then G is positive and Pettis integrable (see [20] (Proposition 2.3)). But according to [20] (Theorem 2.7) G is not McShane integrable. □

Example 3. *Any multifunction G from Example 2 cannot be represented as $G = H + h$, where H is McShane integrable and h is Pettis integrable.*

Proof. If h is a Pettis integrable selection of G, then there exists a measurable function $\alpha : [0,1] \to [0,1]$ such that $h(t) = \alpha(t)g(t)$, for every $t \in [0,1]$.

We have for $H(t) := G(t) - h(t) = conv\{-\alpha(t)g(t), [1-\alpha(t)]g(t)\}$

$$s(x^*, H(t)) = \sup_{0 \le a \le 1} \langle x^*, -a\alpha(t)g(t) + (1-a)[1-\alpha(t)]g(t)\rangle = \sup_{0 \le a \le 1}\langle x^*, g(t)[1-a-\alpha(t)]\rangle$$
$$= \langle x^*, g(t)[1-\alpha(t)]\rangle + \sup_{0 \le a \le 1}\langle x^*, -ag(t)\rangle = \langle x^*, g(t)[1-\alpha(t)]\rangle - \inf_{0 \le a \le 1}\langle x^*, ag(t)\rangle \quad (3)$$
$$= \langle x^*, g(t)[1-\alpha(t)]\rangle + \langle x^*, g(t)\rangle^-$$

If H would be McShane integrable then, the family

$$\{\langle x^*, [1-\alpha(\cdot)]g(\cdot)\rangle + \langle x^*, g(\cdot)\rangle^- : \|x^*\| \le 1\}$$

would be McShane equiintegrable. But in such a case $-H$ is also McShane integrable. Since

$$s(x^*, -H(t)) = \sup_{0 \le a \le 1}\langle x^*, a\alpha(t)g(t) + (1-a)[-1+\alpha(t)]g(t)\rangle$$
$$= -\langle x^*, [1-\alpha(t)]g(t)\rangle + \langle x^*, g(t)\rangle^+ \quad (4)$$

the family $\{\langle x^*, [-1+\alpha(\cdot)]g(\cdot)\rangle + \langle x^*, g(\cdot)\rangle^+ : \|x^*\| \le 1\}$ would be also McShane equiintegrable. Substracting (3) and (4), we obtain McShane equiintegrability of the family

$$\{[1-2\alpha(\cdot)]\langle x^*, g(\cdot)\rangle - \langle x^*, g(\cdot)\rangle \colon \|x^*\| \leq 1\} = \{-2\alpha(\cdot)\langle x^*, g(\cdot)\rangle \colon \|x^*\| \leq 1\}.$$

That means that if H is McShane integrable, then also h is McShane integrable. Consequently, G is McShane integrable, contradicting our assumption. □

Below, we make usage of multifunctions determined by functions, that is the multifunctions of the shape $G(t) = conv\{0, g(t)\}$, where g is a Banach space valued function. We refere to [20], for the relations of integrability between g and G. At this stage we recall only that Henstock integrability of g, in general, does not imply Henstock integrability of G. In fact let g be a Henstock but non McShane integrable function. If, by contradiction, G is Henstock integrable then, by [18] (Proposition 3.1), G is McShane integrable and then, by [20] (Theorem 2.7), g is McShane integrable. For the relations among different types of integrability for vector valued functions we refer also to [51].

Remark 2. *There is now an obvious question: Let $\Gamma \colon [0,1] \to cwk(X)$ be a variationally Henstock (Henstock, $\mathcal{H}$) integrable multifunction. Does there exist a variationally Henstock (Henstock, $\mathcal{H}$) integrable selection f of Γ such that the integral of $G := \Gamma - f$ is of finite variation?*

Unfortunately, in general, the answer is negative. The argument is similar to that applied in [51]. Assume that X is separable and g is the X-valued function constructed in [46] that is vH-integrable (and so it is strongly measurable by [52]) as well as Pettis but not Bochner integrable (see [46]). Let $\Gamma(t) := conv\{0, g(t)\}$. Then, Γ is vH-integrable (see [17] (Example 4.7)) but it is not Bochner integrable because it possesses at least one vH-integrable selection that is not Bochner integrable (see [17] (Theorem 3.7). Let now $f \in \mathcal{S}_{vH}(\Gamma)$ and consider the multifunction $G := \Gamma - f$. Clearly G is vH-integrable (hence also Henstock and $\mathcal{H}$-integrable) and $G(t) = conv\{-f(t), g(t) - f(t)\}$ for all $t \in [0,1]$.

If the integral of G were of finite variation, then G would be Bochner integrable. In fact by Proposition 2, G is Pettis integrable. Since G is compact valued and X is separable, an application of [25] (Proposition 3.5) gives that also $i(G)$ (i is the Rådström embedding) is Pettis integrable. Moreover, since G is Bochner measurable, $i(G)$ is strongly measurable. Now the finite variation of $i(G)$ yields Bochner integrability of $i(G)$. So since G is Bochner measurable it becomes Bochner integrable (an equivalent proof can be deduced from Remark 1). Therefore, the selections $-f, g - f$ would be Bochner integrable since they are strongly measurable and dominated by $\|G\|_h$. But that would mean that g is Bochner integrable, contrary to the assumption.

The multifunction Γ is also an example of a strongly measurable and Birkhoff (McShane) integrable multifunction (see [17] (Theorem 4.3)) that cannot be decomposed into Birkhoff (McShane) integrable multifunction with integral of finite variation and a selection.

Example 4. *There exists a McShane integrable multifunction $G \colon [0,1] \to cwk(X)$ such that $0 \in G(t)$ for every $t \in [0,1]$, but G is not Birkhoff integrable. Moreover, G cannot be represented as $G = H + h$, where $H \colon [0,1] \to cwk(X)$ is Birkhoff integrable and $h \colon [0,1] \to X$ is McShane integrable. G may be chosen with its integral of finite variation.*

Proof. We take in Example 2 a function g that is McShane but not Birkhoff integrable and follow the same calculations. The second assertion can be proved as that in Example 3. If g is bounded, then the variation of the McShane integral of G is finite. Phillips' function is an example of such a function. As proved in [53] (Example 2.1) it is McShane integrable but not Birkhoff. □

Example 5. *Let $X = \ell_2[0,1]$ and let $\{e_t : t \in (0,1]\}$ be its orthonormal basis. Let $G(t) := conv\{0, e_t\}, t \in (0,1]$. Then G is Birkhoff integrable and bounded (cf. [20] (Example 2.11)). G cannot be represented as $G = H + h$, where h is a Birkhoff integrable selection of G and H is Bochner integrable.*

Proof. Suppose that such a representation exists: $G = H + h$. Then there exists a measurable function $\alpha \colon [0,1] \to [0,1]$ such that $h(t) = \alpha(t)e_t$ for all $t \in (0,1]$. We may assume that α is positive on a set of positive Lebesgue measure. Then, $H(t) = conv\{\alpha(t)e_t, (1-\alpha(t))e_t\}$. Since H is - by definition -

Bochner measurable, there exists a set $K \subset [0,1]$ of full measure such that $\{H(t)\colon t \in K\}$ is separable in d_H. But if $t \neq t'$, then
$$d_H(H(t), H(t')) \geq \max\{\alpha(t), \alpha(t')\}.$$

Hence there is $\varepsilon > 0$ such that $d_H(H(t), H(t')) \geq \varepsilon > 0$ on a set of positive measure. However, that contradicts the separability. □

Proposition 3. *Let $G\colon [0,1] \to cwk(X)$ be McShane integrable (hence also Henstock) such that its integral $M_G\colon \mathcal{L} \to cwk(X)$ is of finite variation. If $G := H + h$, where h is a McShane integrable selection of G, then the variation of the multiintegral M_H of H is finite. Moreover H is Birkhoff and variationally Henstock integrable.*

Proof. Let G be McShane integrable and such that $|M_G| < \infty$ (in [53] (Example 2.1) there is an example of such a G that is also not Birkhoff integrable). Let ν_h be the McShane integral of h. Since h is a selection of G, we have $\nu_h(E) \in M_G(E)$ for every $E \in \mathcal{L}$. Consequently $|\nu_h|[0,1] \leq |M_G|[0,1] < \infty$ and then $|M_H|[0,1] \leq |M_G|[0,1] + |\nu_h|[0,1] < \infty$. Moeover by [19] (Corollary 3.7) we get that H is Birkhoff and variationally Henstock integrable. □

Now, to provide the reader with a quick overview of decomposition results which can be derived from Propositions 1 and 2 and from the articles quoted in the list of references, we have collected the results in Table 2A,B for gauge integrals and in Tables 3 and 4 for scalarly defines integrals.

In the left column of the subsequent tables there are multifunctions G of different type. In the first row there are functions f with the corresponding properties. In the intersection of a row α and a column β one finds a class V of multifunctions Γ together with equality or an inclusion.

- The notation $= V$ means that each element of V can be represented as $G + f$, where f is a selection of Γ belonging to the class β and G is a member of the class α. And conversely, if $G \in \alpha$ and $f \in \beta$, then $G + f \in V$.
- The inclusion $\subset V$ means that if $G \in \alpha$ and $f \in \beta$, then $G + f \in V$. While $\subsetneq V$ means that if $G \in \alpha$ and $f \in \beta$, then $G + f \in V$ but there are elements Γ of V that cannot be represented as $\Gamma = G + f$, where $G \in \alpha$ and f is a selection of Γ belonging to β. Clearly, one has always $\Gamma = \Gamma + 0$ but, if zero function is not a selection of Γ then this is not what we are looking for.
- The inclusion $\supset V$ means that each element of V can be represented as $G + f$, where f is a selection of Γ belonging to the class β and G is a member of the class α. While $\supsetneq V$ means additionally that sometimes $G + f \notin V$ for properly chosen G and f.
- Question tag indicates that we do not know something.

In Table 2A,B we describe decomposition into gauge integrable multifunction and function. Similarly as in case of Table 1 Henstock, $\mathcal{H}$ and $v\mathcal{H}$-integrable functions possessing integrals of finite variation, are not taken into consideration, because such functions are McShane and Birkhoff integrable, respectively ([21] (Theorem 4.5)).

In the tables that follow the most significant results will be highlighted by a box.

Table 2. Part A: Decomposition: $\Gamma = G + f$, arbitrary gauge defined G and f. **Part B**: Decomposition: $\Gamma = G + f$, arbitrary gauge defined G and f.

	Gf	L_1	Bi_{fv}	Bi	MS_{fv}
A	L_1	$= L_1$	$\subsetneq Bi_{fv}$ Example 5, [54] (Proposition 4.7)	$\subsetneq Bi$ Example 5, [54] (Proposition 4.7), [51] (Example 2)	$\subsetneq MS_{fv}$ Proposition 3, Example 5, [54] (Proposition 4.7)
	Bi_{fv}	$\subset Bi_{fv}$	$= Bi_{fv}$	$\subsetneq Bi$ Example 5, [54] (Proposition 4.7), [51] (Example 2)	$= MS_{fv}$ Proposition 3, Example 5, [54] (Proposition 4.7)
	Bi	$\subsetneq Bi$ Remark 3	$\subsetneq Bi$ Remark 3	$= Bi$	$\subsetneq MS$ Example 4
	$Bi \cap vH$	$\subset Bi \cap vH$ $\neq Bi \cap vH$?	$\subsetneq H$ Remark 3	$\subset Bi$ $\neq Bi$?	$\subsetneq MS$ Example 4
	MS_{fv}	$\subset MS_{fv}$	$\subset MS_{fv}$ $\neq MS_{fv}$?	$\subset MS$ $\neq MS$?	$= MS_{fv}$
	MS	$\subset MS$ Remark 3	$\subset MS$ Remark 3	$\subset MS$ $\neq MS$?	$\subset MS$ Remark 3

	Gf	MS	H	vH	$vH = vH$
B	L_1	$\subsetneq MS$ Example 5 [54] (Proposition 4.7), [51] (Example 2)	$\subsetneq H$ Example 5 [54] (Proposition 4.7)	$\subsetneq H$ Example 5 [54] (Proposition 4.7)	$\subsetneq vH$ [18] (Remark 5.4)
	Bi_{fv}	$\subsetneq MS$ Example 5, [54] (Proposition 4.7), [51] (Example 2)	$\subsetneq H$ Remark 2	$\subsetneq H$ Remark 2	$\subsetneq H$ Remark 2
	Bi	$\subsetneq MS$ Example 4	$\subsetneq H$ Remark 2	$= H$ [18] (Theorem 4.2)	$\subsetneq H$ & $\supset vH$ [19] (Cor. 3.7)
	$Bi \cap vH$	$\subsetneq MS$ Proposition 3	$\subsetneq H$ Remark 3	$\subsetneq H$ Remark 3	$= vH$ [18] (Theorem 5.3), [19] (Cor. 3.7)
	MS_{fv}	$\subsetneq MS$, [51] (Example 2)	$\subsetneq H$ Remark 2	$\subsetneq H$ Remark 2	$\subsetneq H$ Remark 2
	MS	$= MS$	$= H$ [18] (Theorem 3.2)	$\supsetneq H$ [18] (Theorem 4.2)	$\subset H$

Table 3. $\Gamma = G + f$. G and f scalarly defined.

	G_f	P_e	$P_{e_{fo}}$	wMS	HKP	DP
G_f						
P_e	$=P_e$		$\subseteq Pe$ Remark 3	$=wMS$	$=HKP$, [31] (Theorem 1) [51] (Theorem 1) (sep. case)	$=DP$ [21] (Theorem 3.5)
$P_{e_{fo}}$	$\subseteq P_e$		$=P_{e_{fo}}$	$\subset wMS$	$\subset HKP$	$\subset DP$
wMS	$\subsetneq wMS$ [21] (Theorem 3.7)		$\subsetneq wMS$ [21] (Theorem 3.7)	$= wMS$	$=HKP$ [21] (Theorem 3.5)	$=DP$ [21] (Theorem 3.5)
HKP	$\subsetneq HKP$ [21] (Theorem 3.7)		$\subsetneq HKP$ [21] (Theorem 3.7)	$\subsetneq HKP$ [21] (Theorem 3.7)	$=HKP$	$=DP$ [21] (Theorem 3.5)
DP	$\subsetneq DP$ [21] (Theorem 3.7)		$\subsetneq DP$ [21] (Theorem 3.7)	$\subsetneq DP$ [21] (Theorem 3.7)	$\subsetneq DP$ [21] (Theorem 3.7)	$=DP$

Table 4. $\Gamma = G + f$. Arbitrary G and f.

Gf	Pe	wMS	HKP	DP
Pe + DL	$= P_e +$ DL	$= wMS +$ DL	$=$ HKP + DL	$=$ DP + DL
Pe + Db	$= P_e +$ Db	$= wMS +$ Db	$=$ HKP + Db	$=$ DP + Db

Remark 3. *We observe that*

1. (Bi, H)-cell and $(Bi \cap vH, H)$-cell: Multifunction Γ that is Henstock integrable but not $\mathcal{H}$-integrable cannot be decomposed as $\Gamma = G + f$ with Birkhoff integrable G. G is only McShane integrable.
2. $(Bi \cap vH, \mathcal{H})$-cell: Multifunction Γ that is $\mathcal{H}$-integrable but not vH-integrable cannot be represented as $\Gamma = G + f$ with Birkhoff and vH-integrable G.
3. The Henstock (resp. $\mathcal{H}$) integrability of G, together with $0 \in G(t)$ a.e. implies that G is McShane integrable (resp. Bi) by [18] (Proposition 3.1) and then the characterization any class of Γ is contained in the MS and Bi rows.
4. The vH integrability of G, together with $0 \in G(t)$ a.e. implies that G is Birkhoff integrable by [17] (Theorem 4.1), in particular if the selection f is vH-integrable then we have $vH \ni \Gamma = G + f$ by [18] (Theorem 5.3), or [19] (Cor. 3.7).
5. (MS, MS_{vf})-cell: Let f be McShane integrable with $|\nu_f|[0,1] = +\infty$. Define Γ by $\Gamma(t) = \text{conv}\{f(t)/2, f(t)\}$. The multifunction Γ is McShane integrable and the integral of each scalarly measurable selection of Γ is of infinite variation.
6. (Bi, L_1) and (Bi, Bi_{vf})-cells: The same as in (5) but with a Birkhoff integrable function.

Now we are going to describe decompositions into scalarly integrable multifunctions and functions. In Table 3 there are no multifunctions that are wMS, HKP or DP integrable and their integrals are of finite variation. In virtue of [54] (Theorem 3.2) such multifunctions are Pettis integrable.

If we assume in addition that G satisfies the Db-condition (resp. DL-condition) we are able to find the relations below (cfr. [54] (Theorem 4.1)).

Remark 4. *It seems that the decomposition $\Gamma = G + f$ with $G \in wMS \cup HKP \cup DP$ is useless if Γ is Pettis or stronger integrable. If Γ Henstock, $\mathcal{H}$ or vH integrable, then Table 2A,B give better decompositions. As an example in Table 5 we assume Pettis integrability of G.*

Table 5. Decomposition: $\Gamma = G + f$, scalarly def. G and gauge def. f

Gf	L_1	Bi_{fv}	Bi	MS_{fv}	MS	H	$\mathcal{H}$	$vH = v\mathcal{H}$
P_e	$\supsetneq L_1$	$\supsetneq Bi_{fv}$	$\supsetneq Bi$	$\supsetneq MS_{fv}$	$\supsetneq MS$	$\supsetneq MS$	$\supsetneq Bi$	$\supsetneq vH$
P_{efv}	$\supsetneq L_1$	$\supsetneq Pe_{fv}$	$\supsetneq Bi_{fv}$	$\supsetneq MS_{fv}$	$\supsetneq MS_{fv}$	$\supsetneq MS_{fv}$	$\supsetneq Bi_{fv}$	$\supsetneq L_1$

Remark 5. *One would like to have yet decompositions $\Gamma = G + f$ with gauge defined G and scalarly defined f. Unfortunately, the Pettis row in Table 3 seems to be top of what can be obtained. Pettis integrability seems to be resistant to gauge integrable selections. If f is a Henstock integrable selection of a Pettis integrable $\Gamma : [0,1] \to cwk(X)$, then $\Gamma = G + f$ and f is Pettis integrable. Hence f is McShane integrable and G is Pettis integrable. We are unable to conclude any stronger type integrability for G (see Examples 2 and 3). Therefore, we do not present the corresponding table.*

One could expect that if we assume Bochner measurability of G and strong measurability of f in the above tables, then we should get more information. Unfortunately, the answer is negative. The only positive fact is the equality of Pettis, McShane and Birkhoff integrabilities for multifunctions and functions and Bochner integrability in case of integrals of finite variation. Other interrelations remain exactly the same as in the tables presented above.

Moreover, we want to recall that results on decompositions were also obtained for scalarly defined and gauge integrals in the fuzzy setting, as generalization of the multivalued case, in the papers [55–57].

5. Conclusions

As we wrote in the introduction, a more general theory for the multivalued integration is not sufficiently developped until now. In the particular case of closed convex sets, only some results are known [21]. It should be interesting to also develop the theory in such a more general case.

Author Contributions: The authors contributed jointly to this work, have read and agreed to the published version of the manuscript. All authors have read and agreed to the published version of the manuscript.

Funding: This research was partially supported by Grant "Analisi reale, teoria della misura ed approssimazione per la ricostruzione di immagini" (2020) of GNAMPA—INDAM (Italy) and by University of Palermo—Fondo Ricerca di Base 2019).

Conflicts of Interest: The authors declare no conflict of interest.

References

1. Balder, E.J.; Sambucini, A.R. Fatou's lemma for multifunctions with unbounded values in a dual space. *J. Convex Anal.* **2005**, *12*, 383–395.
2. Cichoń, K.; Cichoń, M. Some Applications of Nonabsolute Integrals in the Theory of Differential Inclusions in Banach Spaces. In *Vector Measures, Integration and Related Topics*; Curbera, G.P., Mockenhaupt, G., Ricker, W.J., Eds.; Operator Theory: Advances and Applications; BirHauser-Verlag: Basel, Switzerland, 2010; Volume 201, pp. 115–124, ISBN 978-3-0346-0210-5.
3. Di Piazza, L.; Marraffa, V.; Satco, B. Set valued integrability and measurability in non separable Fréchet spaces and applications. *Math. Slovaca* **2016**, *66*, 1119–1138. [CrossRef]
4. Di Piazza, L.; Marraffa, V.; Satco, B. Closure properties for integral problems driven by regulated functions via convergence results. *J. Math. Anal. Appl.* **2018**, *466*, 690–710. [CrossRef]
5. Di Piazza, L.; Satco, B. A new result on impulsive differential equations involving non-absolutely convergent integrals. *J. Math. Anal. Appl.* **2009**, *352*, 954–963. [CrossRef]
6. Hu, S.; Papageorgiou, N.S. Handbook of Multivalued Analysis I and II. In *Mathematics and Its Applications*, *419*; Kluwer Academic Publisher: Dordrecht, The Netherlands, 1997.
7. Labuschagne, C.C.A.; Marraffa, V. On spaces of Bochner and Pettis integrable functions and their set-valued counterparts. *Nonlinear Math. Uncertainty Appl. AISC* **2011**, *100*, 51–59.
8. Kudo, H. Dependent experiments and sufficient statistics. *Nat. Sci. Rep. Ochanomizu Univ.* **1954**, *4*, 151–163.
9. Richter, H. Verallgemeinerung eines in der Statistik benötigten Satzes der Masstheorie, (German). *Math. Ann.* **1963**, *150*, 85–90. [CrossRef]
10. Shang, Y. The limit behavior of a stochastic logistic model with individual time-dependent rates. *J. Math.* **2013**, *2013*, 1–8. [CrossRef]
11. Boccuto, A.; Candeloro, D.; Sambucini, A.R. Henstock multivalued integrability in Banach lattices with respect to pointwise non atomic measures. *Atti Accad. Naz. Lincei Rend. Lincei Mat. Appl.* **2015** *26*, 363–383. [CrossRef]
12. Boccuto, A.; Sambucini, A.R. A note on comparison between Birkhoff and McShane-type integrals for multifunctions. *Real Anal. Exch.* **2011**, *37*, 315–324. [CrossRef]
13. Bongiorno, B.; Di Piazza, L.; Musiał, K. A variational Henstock integral characterization of the Radon-Nikodym property. *Ill. J. Math.* **2009**, *53*, 87–99. [CrossRef]
14. Boxer, L. Multivalued Functions in Digital Topology. *Note di Matematica* **2017**, *37*, 61–76. [CrossRef]
15. Candeloro, D.; Croitoru, A.; Gavrilut, A.; Sambucini, A.R. An extension of the Birkhoff integrability for multifunctions. *Mediterr. J. Math.* **2016**, *13*, 2551–2575. [CrossRef]
16. Candeloro, D.; Croitoru, A.; Gavrilut, A.; Iosif, A.; Sambucini, A.R. Properties of the Riemann-Lebesgue integrability in the non-additive case. *Rend. Circ. Mat. Palermo II Ser.* **2019**. [CrossRef]
17. Candeloro, D.; Di Piazza, L.; Musiał, K.; Sambucini, A.R. Gauge integrals and selections of weakly compact valued multifunctions. *J. Math. Anal. Appl.* **2016**, *441*, 293–308. [CrossRef]

18. Candeloro, D.; Di Piazza, L.; Musiał, K.; Sambucini, A.R. Relations among gauge and Pettis integrals for multifunctions with weakly compact convex values. *Annali di Matematica* **2018**, *197*, 171–183. [CrossRef]
19. Candeloro, D.; Di Piazza, L.; Musiał, K.; Sambucini, A.R. Some new results on integration for multifunction. *Ricerche Mat.* **2018**, *67*, 361–372. [CrossRef]
20. Candeloro, D.; Di Piazza, L.; Musiał, K.; Sambucini, A.R. Multifunctions determined by integrable functions. *Inter. J. Approx. Reason.* **2019**, *112*, 140–148. [CrossRef]
21. Candeloro, D.; Di Piazza, L.; Musiał, K.; Sambucini, A.R. Integration of multifunctions with closed convex values in arbitrary Banach spaces. in press *J. Convex Anal.* **2020**, *27*.
22. Candeloro, D.; Sambucini, A.R. A Girsanov result through Birkhoff integral. In *Computational Science and Its Applications ICCSA*; Gervasi, O., Murgante, B., Misra, S., Stankova, E., Torre, C.M., Rocha, A.M.A.C., Taniar, D., Apduhan, B.O., Tarantino, E., Ryu, Y., Eds.; LNCS 10960; Springer: Cham, Switzerland, 2018; pp. 676–683.
23. Candeloro, D.; Sambucini, A.R.; Trastulli, L. A vector Girsanov result and its applications to conditional measures via the Birkhoff integrability. *Mediterr. J. Math.* **2019**, *16*, 144. [CrossRef]
24. Caponetti, D.; Marraffa, V.; Naralenkov, K. On the integration of Riemann-measurable vector-valued functions. *Monatsh. Math.* **2017**, *182*, 513–536. [CrossRef]
25. Cascales, C.; Kadets, V.; Rodríguez, J. Birkhoff integral for multi-valued functions. *J. Math. Anal. Appl.* **2004**, *297*, 540–560. [CrossRef]
26. Cascales, C.; Kadets, V.; Rodríguez, J. Measurable selectors and set-valued Pettis integral in non-separable Banach spaces. *J. Funct. Anal.* **2009**, *256*, 673–699. [CrossRef]
27. Cascales, C.; Kadets, V.; Rodríguez, J. The Gelfand integral for multi-valued functions. *J. Convex Anal.* **2011**, *18*, 873–895.
28. D'Aniello, E.; Mauriello, M. Some Types of Composition Operators on Some Spaces of Functions. *arXiv* **2020**, arXiv:2005.07735.
29. Di Piazza, L.; Marraffa, V.; Musiał, K. Variational Henstock integrability of Banach space valued function. *Math. Bohem.* **2016**, *141*, 287–296. [CrossRef]
30. Di Piazza, L.; Musiał, K. A decomposition theorem for compact-valued Henstock integral. *Monatsh. Math.* **2006**, *148*, 119–126. [CrossRef]
31. Di Piazza, L.; Musiał, K. A decomposition of Denjoy-Khintchine-Pettis and Henstock-Kurzweil-Pettis integrable multifunctions. In *Vector Measures, Integration and Related Topics*; Curbera, G.P., Mockenhaupt, G., Ricker, W.J., Eds.; Operator Theory: Advances and Applications; Birkhauser Verlag: Basel, Switzerland, 2009; Volume 201, pp. 171–182.
32. Di Piazza, L.; Musiał, K. Henstock-Kurzweil-Pettis integrability of compact valued multifunctions with values in an arbitrary Banach space. *J. Math. Anal. Appl.* **2013**, *408*, 452–464. [CrossRef]
33. Di Piazza, L.; Musiał, K. Relations among Henstock, McShane and Pettis integrals for multifunctions with compact convex values. *Monatsh. Math.* **2014**, *173*, 459–470. [CrossRef]
34. Di Piazza, L.; Porcello, G. Radon-Nikodym theorems for finitely additive multimeasures. *Z. Anal. Ihre. Anwend. (ZAA)* **2015**, *34*, 373–389. [CrossRef]
35. Kaliaj, S.B. The New Extensions of the Henstock–Kurzweil and the McShane Integrals of Vector-Valued Functions. *Mediterr. J. Math.* **2018**, *15*, 22. [CrossRef]
36. Musiał, K. Pettis Integrability of Multifunctions with Values in Arbitrary Banach Spaces. *J. Convex Anal.* **2011**, *18*, 769–810.
37. Musiał, K. Approximation of Pettis integrable multifunctions with values in arbitrary Banach spaces. *J. Convex Anal.* **2013**, *20*, 833–870.
38. Naralenkov, K.M. A Lusin type measurability property for vector-valued functions. *J. Math. Anal. Appl.* **2014**, *417*, 293–307. [CrossRef]
39. Gordon, R.A. The Denjoy extension of the Bochner, Pettis and Dunford integrals. *Stud. Math.* **1989**, *92*, 73–91. [CrossRef]
40. Candeloro, D.; Sambucini, A.R. Order-type Henstock and McShane integrals in Banach lattices setting. In Proceedings of the Sisy 20014- IEEE 12th International Symposium on Intelligent Systems and Informatics, Subotica, Serbia, 11–13 September 2014; Volume 9.
41. El Amri K.; Hess, C. On the Pettis integral of closed valued multifunctions. *Set-Valued Anal.* **2000**, *8*, 329–360. [CrossRef]

42. Shang, Y. Continuous-time average consensus under dynamically changing topologies and multiple time-varying delays. *Appl. Math. Comput.* **2014**, *244*, 457–466. [CrossRef]
43. Labuschagne, C.C.A.; Pinchuck, A.L.; van Alten, C.J. A vector lattice version of Rådström's embedding theorem. *Quaest. Math.* **2007**, *30*, 285–308. [CrossRef]
44. Fremlin, D.H. The generalized McShane integral. *Ill. J. Math.* **1995**, *39*, 39–67. [CrossRef]
45. Fremlin, D.H. *The McShane and Birkhoff Integrals of Vector-Valued Functions*; Mathematics Department Research Report 92-10, Version 18.5; University of Essex: Colchester, UK, 2007.
46. Di Piazza, L.; Marraffa, V. The McShane, PU and Henstock integrals of Banach valued functions. *Czechoslov. Math. J.* **2002**, *52*, 609–633. [CrossRef]
47. Musiał, K. Topics in the theory of Pettis integration. *Rend. Istit. Mat. Univ. Trieste* **1991**, *23*, 177–262.
48. Di Piazza, L.; Musiał, K. A characterization of variationally McShane integrable Banach-space valued function. *Ill. J. Math.* **2001**, *45*, 279–289. [CrossRef]
49. Gordon, R.A. The McShane integral of Banach-valued functions. *Ill. J. Math.* **1990**, *34*, 557–567.
50. Saadoune, M.; Sayyad, R. The weak Mc Shane integral. *Czechoslov. Math. J.* **2014**, *64*, 387–418. [CrossRef]
51. Di Piazza, L.; Musiał, K. Set-Valued Kurzweil-Henstock-Pettis Integral. *Set-Valued Anal.* **2005**, *13*, 167–179. [CrossRef]
52. Cao, S. The Henstock integral for Banach-valued functions, The Henstock integral for Banach-valued functions. *SEA Bull. Math.* **1992**, *16*, 35–40.
53. Rodríguez, J. Some examples in Vector Integration. *Bull. Aust. Math. Soc.* **2009**, *80*, 384–392. [CrossRef]
54. Candeloro, D.; Di Piazza, L.; Musiał, K.; Sambucini, A.R. Multi-integrals of finite variation. *Boll. dell'Unione Matematica Italiana* **2020**. [CrossRef]
55. Bongiorno, B.; Di Piazza, L.; Musiał, K. A Decomposition Theorem for the Fuzzy Henstock Integral. *Fuzzy Sets Syst.* **2012**, *200*, 36-47. [CrossRef]
56. Di Piazza, L.; Marraffa, V. Pettis integrability of fuzzy mappings with values in arbitrary Banach spaces, functions. *Math. Slovaca* **2017**, *67*, 1359–1370. [CrossRef]
57. Musiał, K. A decomposition theorem for Banach space valued fuzzy Henstock integral. *Fuzzy Sets Syst.* **2015**, *259*, 21–28. [CrossRef]

© 2020 by the authors. Licensee MDPI, Basel, Switzerland. This article is an open access article distributed under the terms and conditions of the Creative Commons Attribution (CC BY) license (http://creativecommons.org/licenses/by/4.0/).

Article

Kuelbs–Steadman Spaces for Banach Space-Valued Measures

Antonio Boccuto [1,*,†]**, Bipan Hazarika** [2,†] **and Hemanta Kalita** [3,†]

[1] Department of Mathematics and Computer Sciences, University of Perugia, via Vanvitelli, 1 I-06123 Perugia, Italy
[2] Department of Mathematics, Gauhati University, Guwahati 781014, Assam, India; bh_gu@gauhati.ac.in
[3] Department of Mathematics, Patkai Christian College (Autonomous), Dimapur, Patkai 797103, Nagaland, India; hemanta30kalita@gmail.com
* Correspondence: antonio.boccuto@unipg.it
† These authors contributed equally to this work.

Received: 2 June 2020; Accepted: 15 June 2020; Published: 19 June 2020

Abstract: We introduce Kuelbs–Steadman-type spaces (KS^p spaces) for real-valued functions, with respect to countably additive measures, taking values in Banach spaces. We investigate the main properties and embeddings of L^q-type spaces into KS^p spaces, considering both the norm associated with the norm convergence of the involved integrals and that related to the weak convergence of the integrals.

Keywords: Kuelbs–Steadman space; Henstock–Kurzweil integrable function; vector measure; dense embedding; completely continuous embedding; Köthe space; Banach lattice

MSC: Primary 28B05; 46G10; Secondary 46B03; 46B25; 46B40

---

1. Introduction

Kuelbs–Steadman spaces have been the subject of many recent studies (see, e.g., [1–3] and the references therein). The investigation of such spaces arises from the idea to consider the L^1 spaces as embedded in a larger Hilbert space with a smaller norm and containing in a certain sense the Henstock–Kurzweil integrable functions. This allows giving several applications to functional analysis and other branches of mathematics, for instance Gaussian measures (see also [4]), convolution operators, Fourier transforms, Feynman integrals, quantum mechanics, differential equations, and Markov chains (see also [1–3]). This approach allows also developing a theory of functional analysis that includes Sobolev-type spaces, in connection with Kuelbs–Steadman spaces rather than with classical L^p spaces.

Moreover, in recent studies about integration theory, multifunctions have played an important role in applications to several branches of science, like for instance control theory, differential inclusions, game theory, aggregation functions, economics, problems of finding equilibria, and optimization. Since neither the Riemann integral, nor the Lebesgue integral are completely satisfactory concerning the problem of the existence of primitives, different types of integrals extending the previous ones have been introduced and investigated, like Henstock–Kurzweil, McShane, and Pettis integrals. These topics have many connections with measures taking values in abstract spaces, and in particular, the extension of the concept of integrability to set-valued functions can be used in order to obtain a larger number of selections for multifunctions, through their estimates and properties, in several applications (see, e.g., [5–13]).

In this paper, we extend the theory of Kuelbs–Steadman spaces to measures μ defined on a σ-algebra and with values in a Banach space X. We consider an integral for real-valued functions f

with respect to X-valued countably additive measures. In this setting, a fundamental role is played by the separability of μ. This condition is satisfied, for instance, when T is a metrizable separable space, not necessarily with a Schauder basis (such spaces exist; see, for instance, [1]), and μ is a Radon measure. In the literature, some deeply investigated particular cases are when $X = \mathbb{R}^n$ and μ is the Lebesgue measure, and when X is a Banach space with a Schauder basis (see also [1–3]). Since the integral of f with respect to μ is an element of X, in general, it is not natural to define an inner product, when it is dealt with by norm convergence of the involved integrals. Moreover, when μ is a vector measure, the spaces $L^p[\mu]$ do not satisfy all classical properties as the spaces L^p with respect to a scalar measure (see also [14–16]). However, it is always possible to define Kuelbs–Steadman spaces as Banach spaces, which are completions of suitable L^p spaces. We introduce them and prove that they are normed spaces and that the embeddings of $L^q[\mu]$ into $KS^p[\mu]$ are completely continuous and dense. We show that the norm of KS^p spaces is smaller than that related to the space of all Henstock–Kurzweil integrable functions (the Alexiewicz norm). We prove that KS^p spaces are Köthe function spaces and Banach lattices, extending to the setting of $KS^p[\mu]$-spaces some results proven in [16] for spaces of type $L^p[\mu]$. Furthermore, when X' is separable, it is possible to consider a topology associated with the weak convergence of integrals and to define a corresponding norm and an inner product. We introduce the Kuelbs–Steadman spaces related to this norm and prove the analogous properties investigated for KS^p spaces related to norm convergence of the integrals. In this case, since we deal with a separable Hilbert space, it is possible to consider operators like convolution and Fourier transform and to extend the theory developed in [1–3] to the context of Banach space-valued measures.

2. Vector Measures, *(HKL)*- and *(KL)*-Integrals

Let $T \neq \emptyset$ be an abstract set, $\mathcal{P}(T)$ be the class of all subsets of T, $\Sigma \subset \mathcal{P}(T)$ be a σ-algebra, X be a Banach space, and X' be its topological dual. For each $A \in \Sigma$, let us denote by χ_A the characteristic function of A, defined by:

$$\chi_A(t) = \begin{cases} 1 & \text{if } t \in A, \\ 0 & \text{if } t \in T \setminus A. \end{cases}$$

A vector measure is a σ-additive set function $\mu : \Sigma \to X$. By the Orlicz–Pettis theorem (see also [17] (Corollary 1.4)), the σ-additivity of μ is equivalent to the σ-additivity of the scalar-valued set function $x'\mu : A \mapsto x'(\mu(A))$ on Σ for every $x' \in X'$. For the literature on vector measures, see also [14,15,17–21] and the references therein.

The variation $|\mu|$ of μ is defined by setting:

$$|\mu|(A) = \sup\left\{\sum_{i=1}^r \|\mu(A_i)\| : A_i \in \Sigma, i = 1, 2, \ldots, r; A_i \cap A_j = \emptyset \text{ for } i \neq j; \bigcup_{i=1}^r A_i \subset A\right\}.$$

We define the semivariation $\|\mu\|$ of μ by:

$$\|\mu\|(A) = \sup_{x' \in X', \|x'\| \leq 1} |x'\mu|(A). \tag{1}$$

Remark 1. *Observe that* $\|\mu\|(A) < +\infty$ *for all* $A \in \Sigma$ *(see also [17] (Corollary 1.19), [15] (§1)).*

The completion of Σ with respect to $\|\mu\|$ is defined by:

$$\widetilde{\Sigma} = \{A = B \cup N : B \in \Sigma, N \subset M \in \Sigma \text{ with } \|\mu\|(M) = 0\}. \tag{2}$$

A function $f : T \to \mathbb{R}$ is said to be μ-measurable if:

$$f^{-1}(B) \cap \{t \in T : f(t) \neq 0\} \in \widetilde{\Sigma}$$

for each Borel subset $B \subset \mathbb{R}$.

Observe that from (1) and (2), it follows that every μ-measurable real-valued function is also $x'\mu$-measurable for every $x' \in X'$. Moreover, it is readily seen that every Σ-measurable real-valued function is also μ-measurable.

We say that μ is Σ-separable (or separable) if there is a countable family $\mathbb{B} = (B_k)_k$ in Σ such that, for each $A \in \Sigma$ and $\varepsilon > 0$, there is $k_0 \in \mathbb{N}$ such that:

$$\|\mu\|(A \Delta B_{k_0}) = \sup_{x' \in X', \|x'\| \leq 1} [|x'\mu|(A \Delta B_{k_0})] \leq \varepsilon \qquad (3)$$

(see also [22]). Such a family $\mathbb{B}$ is said to be μ-dense.

Observe that μ is separable if and only if Σ is μ-essentially countably generated, namely there is a countably generated σ-algebra $\Sigma_0 \subset \Sigma$ such that for each $A \in \Sigma$, there is $B \in \Sigma_0$ with $\mu(A \Delta B) = 0$. The separability of μ is satisfied, for instance, when T is a separable metrizable space, Σ is the Borel σ-algebra of the Borel subsets of T, and μ is a Radon measure (see also [23] (Theorem 4.13), [24] (Theorem 1.0), [19] (§1.3 and §2.6), and [22] (Propositions 1A and 3)).

From now on, we assume that μ is separable, and $\mathbb{B} = (B_k)_k$ is a μ-dense family in Σ with:

$$\|\mu\|(B_k) \leq M = \|\mu\|(T) + 1 \quad \text{for all } k \in \mathbb{N}. \qquad (4)$$

Now, we recall the Henstock–Kurzweil (HK) integral for real-valued functions, defined on abstract sets, with respect to (possibly infinite) non-negative measures. For the related literature, see also [5–13,25–33] and the references therein. When we deal with the (HK)-integral, we assume that T is a compact topological space and Σ is the σ-algebra of all Borel subsets of T. We will not use these assumptions to prove the results, which do not involve the (HK)-integral.

Let $\nu : \Sigma \to \mathbb{R} \cup \{+\infty\}$ be a σ-additive non-negative measure. A decomposition of a set $A \in \Sigma$ is a finite collection $\{(A_1, \xi_1), (A_2, \xi_2), \ldots, (A_N, \xi_N)\}$ such that $A_j \in \Sigma$ and $\xi_j \in A_j$ for every $j \in \{1, 2, \ldots, N\}$, and $\nu(A_i \cap A_j) = 0$ whenever $i \neq j$. A decomposition of subsets of $A \in \Sigma$ is called a partition of A when $\bigcup_{j=1}^{N} A_j = A$. A gauge on a set $A \in \Sigma$ is a map δ assigning to each point $x \in A$ a neighborhood $\delta(x)$ of x. If $\mathcal{D} = \{(A_1, \xi_1), (A_2, \xi_2), \ldots, (A_N, \xi_N)\}$ is a decomposition of A and δ is a gauge on A, then we say that $\mathcal{D}$ is δ-fine if $A_j \subset \delta(\xi_j)$ for any $j \in \{1, 2, \ldots, N\}$.

An example is when T_0 is a locally compact and Hausdorff topological space and $T = T_0 \cup \{x_0\}$ is the one-point compactification of T_0. In this case, we will suppose that all involved functions f vanish on x_0. For instance, this is the case when $T_0 = \mathbb{R}^n$ is endowed with the usual topology and x_0 is a point "at the infinity", or when T is the unbounded interval $[a, +\infty] = [a, +\infty) \cup \{+\infty\}$ of the extended real line, considered as the one-point compactification of the locally compact space $[a, +\infty)$. In this last case, the base of open sets consists of the open subsets of $[a, +\infty)$ and the sets of the type $(b, +\infty]$, where $a < b < +\infty$. Any gauge in $[a, +\infty]$ has the form $\delta(x) = (x - d(x), x + d(x))$, if $x \in [a, +\infty] \cap \mathbb{R}$, and $\delta(+\infty) = (b, +\infty] = (b, +\infty) \cup \{+\infty\}$, where d denotes a positive real-valued function defined on $[a, +\infty)$. Now, we define the Riemann sums by: $S(f, \mathcal{D}) = \sum_{j=1}^{N} f(\xi_j) \nu(A_j)$ if the sum exists in $\mathbb{R}$, with the convention $0 \cdot (+\infty) = 0$. Note that for any gauge δ, there exists at least one δ-fine partition $\mathcal{D}$ such that $S(f, \mathcal{D})$ is well defined.

A function $f: T \to \mathbb{R}$ is said to be Henstock–Kurzweil integrable ((HK)-integrable) on a set $A \in \Sigma$ if there is an element $I_A \in \mathbb{R}$ such that for every $\varepsilon > 0$, there is a gauge δ on A with $|S(f, \mathcal{D}) - I_A| \le \varepsilon$ whenever $\mathcal{D}$ is a δ-fine partition of A such that $S(f, \mathcal{D})$ exists in $\mathbb{R}$, and we write:

$$(HK) \int_A f \, dv = I_A.$$

Observe that, if $A, B \in \Sigma$, $B \subset A$, and $f: T \to \mathbb{R}$ is (HK)-integrable on A, then f is also (HK)-integrable on B and on $A \setminus B$, and:

$$(HK) \int_A f(t) \, dv = (HK) \int_B f(t) \, dv + (HK) \int_{A \setminus B} f(t) \, dv \tag{5}$$

(see also [25] (Propositions 5.14 and 5.15), [33] (Lemma 1.10 and Proposition 1.11)). From (5) used with $A = T$ and $\chi_B f$ instead of f, it follows that, if f is (HK)-integrable on T and $B \in \Sigma$, then:

$$(HK) \int_T \chi_B(t) f(t) \, dv = (HK) \int_B f(t) \, dv.$$

We say that a Σ-measurable function $f: T \to \mathbb{R}$ is Kluvánek–Lewis–Lebesgue μ-integrable, (KL) μ-integrable (resp. Kluvánek–Lewis–Henstock–Kurzweil μ-integrable, or (HKL) μ-integrable) if the following properties hold:

$$f \text{ is } |x'\mu|\text{-Lebesgue (resp. } |x'\mu|\text{-Henstock–Kurzweil) integrable for each } x' \in X', \tag{6}$$

and for every $A \in \Sigma$, there is $x_A^{(L)}$ (resp. $x_A^{(HK)}$) $\in X$ with:

$$x'(x_A^{(L)}) = (L) \int_A f \, d|x'\mu| \quad (\text{resp. } x'(x_A^{(HK)}) = (HK) \int_A f \, d|x'\mu|) \text{ for all } x' \in X', \tag{7}$$

where the symbols (L) and (HK) in (7) denote the usual Lebesgue (resp. Henstock–Kurzweil) integral of a real-valued function with respect to an (extended) real-valued measure. A Σ-measurable function $f: T \to \mathbb{R}$ is said to be weakly (KL) (resp. weakly (HKL)) μ-integrable if it satisfies only condition (6) (see also [18,21,34]). We recall the following facts about the (KL)-integral.

Proposition 1. (See also [21] (Theorem 2.1.5 (i))) *If* $s: T \to \mathbb{R}$, $s = \sum_{i=1}^{r} \alpha_i \chi_{A_i}$ *is Σ-simple, with $\alpha_i \in \mathbb{R}$, $A_i \in \Sigma$, $i = 1, 2, \ldots, r$ and $A_i \cap A_j = \emptyset$ for $i \ne j$, then s is (KL) μ-integrable on T, and:*

$$(KL) \int_A s \, d\mu = \sum_{i=1}^{r} \alpha_i \, \mu(A \cap A_i) \text{ for all } A \in \Sigma.$$

Proposition 2. (See also [21] (Theorem 2.1.5 (vi))) *If* $f: T \to \mathbb{R}$ *is (KL) μ-integrable on T and $A \in \Sigma$, then $\chi_A f$ is (KL) μ-integrable on T and:*

$$(KL) \int_A f \, d\mu = (KL) \int_T \chi_A f \, d\mu.$$

The space $L^1[\mu]$ (resp. $L_w^1[\mu]$) is the space of all (equivalence classes of) (KL) μ-integrable functions (resp. weakly (KL) μ-integrable functions) up to the complement of μ almost everywhere sets. For $p > 1$, the space $L^p[\mu]$ (resp. $L_w^p[\mu]$) is the space of all (equivalence classes of) Σ-measurable functions f such that $|f|^p$ belongs to $L^1[\mu]$ (resp. $L^1_w[\mu]$). The space $L^\infty[\mu]$ is the space of all (equivalence classes of) μ-essentially bounded functions. The norms are defined by:

$$\begin{cases} \|f\|_{L^p[\mu]} = \|f\|_{L^p_w[\mu]} = \sup_{x' \in X', \|x'\| \leq 1} \left((L)\int_T |f(t)|^p \, d|x'\mu|\right)^{1/p} & \text{if } 1 \leq p < \infty, \\ \|f\|_{L^\infty[\mu]} = \sup_{x' \in X', \|x'\| \leq 1} (|x'\mu|\text{-ess sup}|f|) & \end{cases}$$

(see also [35–37]).

If $f : T \to \mathbb{R}$ is an (HKL)-integrable function, then the Alexiewicz norm of f is defined by:

$$\|f\|_{HKL} = \sup_{x' \in X', \|x'\| \leq 1} \left(\sup_{A \in \Sigma} \left|(HK)\int_A f(t) \, d|x'\mu|\right|\right)$$

(see also [38,39]). Observe that, by arguing analogously as in [30] (Theorem 9.5) and [40] (Example 3.1.1), for each $x' \in X'$, we get that $f = 0 \ |x'\mu|$, almost everywhere if and only if $(HK)\int_A f(t) \, d|x'\mu| = 0$ for every $A \in \Sigma$. Thus, it is not difficult to see that $\|\cdot\|_{HKL}$ is a norm. In general, the space of the real-valued Henstock–Kurzweil integrable functions endowed with the Alexiewicz norm is not complete (see also [39] (Example 7.1)).

3. Construction of the Kuelbs–Steadman Spaces and Main Properties

We begin with giving the following technical results, which will be useful later.

Proposition 3. Let $(a_k)_k$ and $(\eta_k)_k$ be two sequences of non-negative real numbers, such that $a = \sup_k a_k < +\infty$, and

$$\sum_{k=1}^\infty \eta_k = 1, \tag{8}$$

and $p > 0$ be a fixed real number. Then,

$$\left(\sum_{k=1}^\infty \eta_k a_k^p\right)^{1/p} \leq a. \tag{9}$$

Proof. We have $\eta_k a_k^p \leq a^p \eta_k$ for all $k \in \mathbb{N}$, and hence:

$$\sum_{k=1}^\infty \eta_k a_k^p \leq a^p \sum_{k=1}^\infty \eta_k = a^p,$$

getting (9). □

Proposition 4. Let $(b_k)_k$, $(c_k)_k$ be two sequences of real numbers, $(\eta_k)_k$ be a sequence of positive real numbers, satisfying (8), and $p \geq 1$ be a fixed real number. Then,

$$\left(\sum_{k=1}^\infty \eta_k |b_k + c_k|^p\right)^{1/p} \leq \left(\sum_{k=1}^\infty \eta_k (|b_k| + |c_k|)^p\right)^{1/p} \leq \left(\sum_{k=1}^\infty \eta_k |b_k|^p\right)^{1/p} + \left(\sum_{k=1}^\infty \eta_k |c_k|^p\right)^{1/p}. \tag{10}$$

Proof. It is a consequence of Minkowski's inequality (see also [41] (Theorem 2.11.24)). □

Let $\mathbb{B} = (B_k)_k$ be as in (4), and set $\mathcal{E}_k = \chi_{B_k}$, $k \in \mathbb{N}$.

For $1 \leq p \leq \infty$, let us define a norm on $L^1[\mu]$ by setting:

$$\|f\|_{KS^p[\mu]} = \begin{cases} \sup\limits_{x' \in X', \|x'\| \leq 1} \left\{ \left[\sum\limits_{k=1}^{\infty} \eta_k \left| (L) \int_T \mathcal{E}_k(t) f(t) d|x'\mu| \right|^p \right]^{1/p} \right\} & \text{if } 1 \leq p < \infty, \\ \sup\limits_{x' \in X', \|x'\| \leq 1} \left[\sup\limits_{k \in \mathbb{N}} \left| (L) \int_T \mathcal{E}_k(t) f(t) d|x'\mu| \right| \right] & \text{if } p = \infty. \end{cases} \quad (11)$$

The following inequality holds.

Proposition 5. *For any $f \in L^1[\mu]$ and $p \geq 1$, it is:*

$$\|f\|_{KS^p[\mu]} \leq \|f\|_{KS^\infty[\mu]}. \quad (12)$$

Proof. By (9) used with:

$$a_k = \left| (L) \int_T \mathcal{E}_k(t) f(t) d|x'\mu|(t) \right|, \quad (13)$$

where x' is a fixed element of X' with $\|x'\| \leq 1$, we have:

$$\left(\sum_{k=1}^{\infty} \eta_k \left| (L) \int_T \mathcal{E}_k(t) f(t) d|x'\mu|(t) \right|^p \right)^{1/p} \leq \sup_{k \in \mathbb{N}} \left| (L) \int_T \mathcal{E}_k(t) f(t) d|x'\mu| \right|. \quad (14)$$

Taking the supremum in (14) as $x' \in X'$, $\|x'\| \leq 1$, we obtain:

$$\|f\|_{KS^p[\mu]} = \sup_{x' \in X', \|x'\| \leq 1} \left\{ \left[\sum_{k=1}^{\infty} \eta_k \left| (L) \int_T \mathcal{E}_k(t) f(t) d|x'\mu| \right|^p \right]^{1/p} \right\}$$
$$\leq \sup_{x' \in X', \|x'\| \leq 1} \left[\sup_{k \in \mathbb{N}} \left| (L) \int_T \mathcal{E}_k(t) f(t) d|x'\mu| \right| \right] = \|f\|_{KS^\infty[\mu]},$$

getting the assertion. □

Now, we prove that:

Theorem 1. *The map $f \mapsto \|f\|_{KS^p[\mu]}$ defined in (11) is a norm.*

Proof. Observe that, by definition, $\|f\|_{KS^p[\mu]} \geq 0$ for every $f \in L^1[\mu]$. Let $f \in L^1[\mu]$ with $\|f\|_{KS^p[\mu]} = 0$. We prove that $f = 0$ μ, almost everywhere. It is enough to take $1 \leq p < \infty$, since the case $p = \infty$ will follow from (12). For $k \in \mathbb{N}$, let a_k be as in (13). As the η_k's are strictly positive, from:

$$\left(\sum_{k=1}^{\infty} \eta_k a_k^p \right)^{1/p} = 0$$

it follows that $a_k = 0$ for every $k \in \mathbb{N}$. Hence,

$$\left| (L) \int_T \mathcal{E}_k(t) f(t) d|x'\mu|(t) \right| = 0 \quad \text{for each } k \in \mathbb{N} \text{ and } x' \in X' \text{ with } \|x'\| \leq 1. \quad (15)$$

Proceeding by contradiction, suppose that $f \neq 0$ μ, almost everywhere. If $E^+ = f^{-1}(]0, +\infty[)$, $E^- = f^{-1}(]-\infty, 0[)$, then $E^+, E^- \in \Sigma$, since f is Σ-measurable, and we have $\mu(E^+) \neq 0$ or $\mu(E^-) \neq 0$.

Suppose that $\mu(E^+) \neq 0$. By the Hahn–Banach theorem, there is $x'_0 \in X'$ with $\|x'_0\| \leq 1$, $x'_0 \mu(E^+) \neq 0$, and hence, $|x'_0 \mu(E^+)| > 0$. Moreover, if $f^*(t) = \min\{f(t), 1\}$, $t \in T$, then $E^+ = \{t \in T : f^*(t) > 0\}$. For each $n \in \mathbb{N}$, set:

$$E_n^+ = \left\{ t \in T : \frac{1}{n+1} < f^*(t) \leq \frac{1}{n} \right\}.$$

Since $E^+ = \bigcup_{n=1}^{\infty} E_n^+$ and $x'_0 \mu$ is σ-additive, there is $\overline{n} \in \mathbb{N}$ with $|x'_0 \mu|(E_{\overline{n}}^+) > 0$. Put $\overline{B} = E_{\overline{n}}^+$, and choose $\overline{\varepsilon}$ such that:

$$0 < \overline{\varepsilon} < \min\left\{ \frac{1}{\overline{n}+1} |x'_0 \mu|(\overline{B}), 1 \right\}. \tag{16}$$

By the separability of μ, in correspondence with $\overline{\varepsilon}$ and $\overline{B}$, there is $B_{k_0} \in \mathbb{B}$ satisfying (3), that is:

$$\|\mu\|(\overline{B} \Delta B_{k_0}) = \sup_{x' \in X', \|x'\| \leq 1} [|x' \mu|(\overline{B} \Delta B_{k_0})] \leq \overline{\varepsilon}. \tag{17}$$

From (16) and (17), we deduce:

$$\|\mu\|(B_{k_0}) \leq \|\mu\|(\overline{B}) + \|\mu\|(\overline{B} \Delta B_{k_0}) < \|\mu\|(T) + 1 = M,$$

so that $B_{k_0} \in \mathbb{B}$, and:

$$\begin{aligned}
\left| (L) \int_T \chi_{B_{k_0}}(t) f(t) \, d|x'_0 \mu|(t) \right| &\geq (L) \int_T \mathcal{E}_{k_0}(t) f(t) \, d|x'_0 \mu|(t) \\
&= (L) \int_{B_{k_0}} f(t) \, d|x'_0 \mu|(t) \geq (L) \int_{B_{k_0}} f^*(t) \, d|x'_0 \mu|(t) \\
&\geq (L) \int_{\overline{B}} f^*(t) \, d|x'_0 \mu|(t) - (L) \int_{\overline{B} \Delta B_{k_0}} f^*(t) \, d|x'_0 \mu|(t) \quad (18) \\
&\geq \frac{1}{\overline{n}+1} |x'_0 \mu|(\overline{B}) - |x' \mu|(\overline{B} \Delta B_{k_0}) \geq \frac{1}{\overline{n}+1} |x'_0 \mu|(\overline{B}) - \overline{\varepsilon} > 0,
\end{aligned}$$

which contradicts (15). Therefore, $\mu(E^+) = 0$.

Now, suppose that $\mu(E^-) \neq 0$. By proceeding analogously as in (18), replacing f with $-f$ and f^* with the function f_* defined by $f_*(t) = \min\{-f(t), 1\}$, $t \in T$, we find an $x'_1 \in X'$ with $\|x'_1\| \leq 1$, an $\overline{\overline{n}} \in \mathbb{N}$, a $\overline{\overline{B}} \in \Sigma$, an $\overline{\overline{\varepsilon}} > 0$, and a $B_{k_1} \in \mathbb{B}$ with $\|\mu\|(B_{k_1}) < M$, and:

$$\left| (L) \int_T \chi_{B_{k_1}}(t) f(t) \, d|x'_1 \mu|(t) \right| \geq (L) \int_{R_{k_1}} f_*(t) \, d|x'_1 \mu|(t) \geq \frac{1}{\overline{\overline{n}}+1} |x'_1 \mu|(\overline{\overline{B}}) - \overline{\overline{\varepsilon}} > 0,$$

getting again a contradiction with (15). Thus, $\mu(E^-) = 0$, and $f = 0$ almost everywhere.

The triangular property of the norm can be deduced from Proposition 4 for $1 \leq p < \infty$, and it is not difficult to see for $p = \infty$; the other properties are easy to check. □

For $1 \leq p \leq \infty$, the Kuelbs-Steadman space $KS^p[\mu]$ (resp. $KS_w^p[\mu]$) is the completion of $L^1[\mu]$ (resp. $L_w^1[\mu]$) with respect to the norm defined in (11) (see also [2–4,35–37]). Observe that, to avoid ambiguity, we take the completion of $L^1[\mu]$ rather than that of $L^p[\mu]$, but since the embeddings in Theorem 2 are continuous and dense, the two methods are substantially equivalent.

By proceeding similarly as in [2] (Theorem 3.26), we prove the following relations between the spaces $L^q[\mu]$ and $KS^p[\mu]$.

Theorem 2. *For every p, q with $1 \leq p \leq \infty$ and $1 \leq q \leq \infty$, it is $L^q[\mu] \subset KS^p[\mu]$ continuously and densely. Moreover, the space of all Σ-simple functions is dense in $KS^p[\mu]$.*

Proof. We first consider the case $1 \leq p < \infty$. Let $f \in L^q[\mu]$, with $1 \leq q < \infty$, and M be as in (4). Note that $M^{\frac{q-1}{q}} \leq M$, since $M \geq 1$. As $|\mathcal{E}_k(t)| = \mathcal{E}_k(t) \leq 1$ and $|\mathcal{E}_k(t)|^q \leq \mathcal{E}_k(t)$ for any $k \in \mathbb{N}$ and $t \in T$, taking into account (9) and applying Jensen's inequality to the function $t \mapsto |t|^q$ (see also [23] (Exercise 4.9)), we deduce:

$$\begin{aligned}
\|f\|_{KS^p[\mu]} &= \sup_{x' \in X', \|x'\| \leq 1} \left\{ \left[\sum_{k=1}^{\infty} \eta_k \left| (L) \int_T \mathcal{E}_k(t) f(t) d|x'\mu| \right|^{\frac{pq}{q}} \right]^{1/p} \right\} \\
&\leq \sup_{x' \in X', \|x'\| \leq 1} \left\{ \left[\sum_{k=1}^{\infty} \eta_k \left((|x'\mu|(B_k))^{q-1} \cdot (L) \int_T \mathcal{E}_k(t) |f(t)|^q d|x'\mu| \right)^{p/q} \right]^{1/p} \right\} \quad (19) \\
&\leq M^{\frac{q-1}{q}} \sup_{x' \in X', \|x'\| \leq 1} \left[\sup_{k \in \mathbb{N}} \left((L) \int_T \mathcal{E}_k(t) |f(t)|^q d|x'\mu| \right)^{1/q} \right] \\
&\leq M \sup_{x' \in X', \|x'\| \leq 1} \left[\left((L) \int_T |f(t)|^q d|x'\mu| \right)^{1/q} \right] = M \|f\|_{L^q[\mu]},
\end{aligned}$$

where M is as in (4). Now, let $1 \leq p < \infty$ and $q = \infty$. We have:

$$\begin{aligned}
\|f\|_{KS^p[\mu]} &= \sup_{x' \in X', \|x'\| \leq 1} \left\{ \left[\sum_{k=1}^{\infty} \eta_k \left| (L) \int_T \mathcal{E}_k(t) f(t) d|x'\mu| \right|^p \right]^{1/p} \right\} \\
&\leq \sup_{x' \in X', \|x'\| \leq 1} \left[(|x'\mu|(B_k))^p \cdot \operatorname{ess\,sup} |f|^p \right]^{1/p} \leq M \cdot \|f\|_{L^\infty[\mu]}. \quad (20)
\end{aligned}$$

The proof of the case $p = \infty$ is analogous to that of the case $1 \leq p < \infty$. Therefore, $f \in KS^p[\mu]$, and the embeddings in (19) and (20) are continuous.

Moreover, observe that every Σ-simple function belongs to $L^q[\mu]$, and the space of all Σ-simple functions is dense in $L^1[\mu]$ with respect to $\|\cdot\|_{L^1[\mu]}$ (see also [21] (Corollary 2.1.10)). Moreover, since $KS^p[\mu]$ is the completion of $L^1[\mu]$ with respect to the norm $\|\cdot\|_{KS^p[\mu]}$, the space $L^1[\mu]$ is dense in $KS^p[\mu]$ with respect to the norm $\|\cdot\|_{KS^p[\mu]}$ (see also [42] (§4.4)).

Choose arbitrarily $\varepsilon > 0$ and $f \in KS^p[\mu]$. There is $g \in L^1[\mu]$ with $\|g - f\|_{KS^p[\mu]} \leq \frac{\varepsilon}{M+1}$. Moreover, in correspondence with ε and g, we find a Σ-simple function s, with $\|s - g\|_{L^1[\mu]} \leq \frac{\varepsilon}{M+1}$. By (19) and (20), $\|\cdot\|_{KS^p[\mu]} \leq M \|\cdot\|_{L^1[\mu]}$, and hence, we obtain:

$$\begin{aligned}
\|s - f\|_{KS^p[\mu]} &\leq \|s - g\|_{KS^p[\mu]} + \|g - f\|_{KS^p[\mu]} \\
&\leq M \|s - g\|_{L^1[\mu]} + \|g - f\|_{KS^p[\mu]} \leq \frac{M\varepsilon}{M+1} + \frac{\varepsilon}{M+1} = \varepsilon,
\end{aligned}$$

getting the last part of the assertion. Thus, the embeddings in (19) and (20) are dense. □

Proposition 6. $KS^\infty[\mu] \subset KS^p[\mu]$ *for every $p \geq 1$.*

Proof. The assertion follows from (12), since $KS^p[\mu]$ (resp. $KS^\infty[\mu]$) is the completion of $L^1[\mu]$ with respect to $\|f\|_{KS^p[\mu]}$ (resp. $\|f\|_{KS^\infty[\mu]}$). □

Remark 2. (a) Notice that, for $q \neq \infty$, by Theorem 2 and Proposition 6, this holds also when $L^q[\mu]$ and $KS^p[\mu]$ are replaced by $L^q_w[\mu]$ and $KS^p_w[\mu]$, respectively.

(b) If f is (HKL)-integrable, then for each $x' \in X'$ and $k \in \mathbb{N}$, $\mathcal{E}_k f$ is both Henstock–Kurzweil and Lebesgue integrable with respect to $|x'\mu|$, since f is Σ-measurable, and the two integrals coincide, thanks to the (HK)-integrability of the characteristic function χ_E for each $E \in \Sigma$ and the monotone convergence theorem (see also [25,33]). Thus, taking into account (14), for every p with $1 \leq p < \infty$, we have:

$$\sup_{x' \in X', \|x'\| \leq 1} \left[\left(\sum_{k=1}^{\infty} \eta_k \left| (L) \int_T \mathcal{E}_k(t) f(t) \, d|x'\mu| \right|^p \right)^{1/p} \right] \leq \sup_{x' \in X', \|x'\| \leq 1} \left(\sup_{k \in \mathbb{N}} \left| (L) \int_T \mathcal{E}_k(t) f(t) \, d|x'\mu| \right| \right)$$

$$= \sup_{x' \in X', \|x'\| \leq 1} \left(\sup_{k \in \mathbb{N}} \left| (HK) \int_T \mathcal{E}_k(t) f(t) \, d|x'\mu| \right| \right) = \sup_{x' \in X', \|x'\| \leq 1} \left(\sup_{k \in \mathbb{N}} \left| (HK) \int_{B_k} f(t) \, d|x'\mu| \right| \right)$$

$$\leq \sup_{x' \in X', \|x'\| \leq 1} \left(\sup_{A \in \Sigma} \left| (HK) \int_A f(t) \, d|x'\mu| \right| \right) = \|f\|_{HKL}. \quad \square$$

The next result deals with the separability of Kuelbs–Steadman spaces, which holds even for $p = \infty$, differently from L^p spaces.

Proposition 7. *For $1 \leq p \leq \infty$, the space $KS^p[\mu]$ is separable.*

Proof. Observe that, by our assumptions, μ is separable, and this is equivalent to the separability of the spaces $L^p[\mu]$ for all $1 \leq p < \infty$ (see also [35] (Proposition 2.3), [22] (Propositions 1A and 3)).

Now, let $\mathcal{H} = \{h_n : n \in \mathbb{N}\}$ be a countable subset of L^1, dense in $L^1[\mu]$ with respect to the norm $\|\cdot\|_{L^1[\mu]}$. By Theorem 2, $\mathcal{H} \subset KS^p[\mu]$. We claim that $\mathcal{H}$ is dense in $KS^p[\mu]$. Pick arbitrarily $\varepsilon > 0$ and $f \in KS^p[\mu]$. There is $g \in L^1[\mu]$ with $\|g - f\|_{KS^p[\mu]} \leq \dfrac{\varepsilon}{M+1}$. In correspondence with ε and g, there exists $n_0 \in \mathbb{N}$ such that $\|h_{n_0} - g\|_{L^1[\mu]} \leq \dfrac{\varepsilon}{M+1}$. By (19), $\|\cdot\|_{KS^p[\mu]} \leq M \|\cdot\|_{L^1[\mu]}$, and hence:

$$\|h_{n_0} - f\|_{KS^p[\mu]} \leq \|h_{n_0} - g\|_{KS^p[\mu]} + \|g - f\|_{KS^p[\mu]} \leq M \|h_{n_0} - g\|_{L^1[\mu]} + \|g - f\|_{KS^p[\mu]}$$

$$\leq \frac{M\varepsilon}{M+1} + \frac{\varepsilon}{M+1} = \varepsilon,$$

getting the claim. $\square$

Now, we prove the following.

Theorem 3. *For $1 \leq p, q < \infty$, the embeddings in (19) are completely continuous, namely map weakly convergent sequences in $L^q[\mu]$ into norm convergent sequences in $KS^p[\mu]$.*

Proof. Pick arbitrarily $1 \leq q < \infty$, and let $(f_n)_n$ be a sequence of elements of $L^q[\mu]$, weakly convergent in $L^q[\mu]$. Then, we get:

$$V = \sup_{n \in \mathbb{N}} \|f_n - f\|_{L^q[\mu]} < +\infty \tag{21}$$

(see also [23] (Proposition 3.5 (iii))) and:

$$\lim_{n \to +\infty} (KL) \int_T \chi_A(t)(f_n(t) - f(t)) \, d\mu = 0 \quad \text{for every } A \in \Sigma \tag{22}$$

(see also [14,15]). Now, let us consider the family of operators $W_k : L^q[\mu] \to X, k \in \mathbb{N}$, defined by:

$$W_k(g) = (KL) \int_T \mathcal{E}_k(t) g(t) \, d\mu, \quad g \in L^q[\mu].$$

It is not difficult to check that W_k is well defined and is a linear operator for every $k \in \mathbb{N}$. Moreover, since $0 \leq \mathcal{E}_k(t) \leq 1$ for all $k \in \mathbb{N}$ and $t \in T$ and taking into account [21] (Theorem 2.1.5 (iii)), for every $g \in L^q[\mu]$, we get:

$$\sup_{x' \in X', \|x'\| \leq 1} \left| (L) \int_T \mathcal{E}_k(t) g(t) \, d|x'\mu| \right|^q \leq \sup_{x' \in X', \|x'\| \leq 1} \left((L) \int_T |g(t)|^q \, d|x'\mu| \right) = \|g\|_{L^q[\mu]}^q < +\infty, \quad (23)$$

and hence, $\sup_k \|W_k(g)\|_X < +\infty$. From (23), it follows also that W_k is a continuous operator for every $k \in \mathbb{N}$. From (21) and the uniform boundedness principle, we deduce:

$$+\infty > W = \sup_{k,n} \|W_k(f_n - f)\|_X = \sup_{x' \in X', \|x'\| \leq 1} \left(\sup_{k,n} \left| (L) \int_T \mathcal{E}_k(t) (f_n(t) - f(t)) \, d|x'\mu| \right| \right). \quad (24)$$

Now, choose arbitrarily $\varepsilon > 0$ and $1 \leq p < \infty$. Note that, by Theorem 2, $f, f_n \in KS^p[\mu]$ for all $n \in \mathbb{N}$. By arguing similarly as in [14] (Appendix 2.3), we find a positive integer K_0 such that $\sum_{k=K_0+1}^{\infty} \eta_k \leq \varepsilon$. Taking into account (9), from (24), we obtain:

$$\sum_{k=K_0+1}^{\infty} \eta_k \left| (L) \int_T \mathcal{E}_k(t)(f_n(t) - f(t)) d|x'\mu| \right|^p \leq \varepsilon W^p \quad (25)$$

for each $n \in \mathbb{N}$ and $x' \in X'$ with $\|x'\| \leq 1$. Moreover, by (22) used with $A = B_k$, $k = 1, 2, \ldots, K_0$, we find a positive integer n^* with:

$$\sum_{k=1}^{K_0} \eta_k \left| (L) \int_T \mathcal{E}_k(t)(f_n(t) - f(t)) d|x'\mu| \right|^p \leq \varepsilon \quad (26)$$

whenever $n \geq n^*$ and $x' \in X'$, $\|x'\| \leq 1$. From (25) and (26), we obtain:

$$\|f_n - f\|_{KS^p[\mu]} = \sup_{x' \in X', \|x'\| \leq 1} \left\{ \left[\sum_{k=1}^{\infty} \eta_k \left| (L) \int_T \mathcal{E}_k(t)(f_n(t) - f(t)) d|x'\mu| \right|^p \right]^{1/p} \right\} \leq \varepsilon^{1/p} (1 + W^p)^{1/p}$$

for all $n \geq n^*$. Thus, the sequence $(f_n)_n$ norm converges in $KS^p[\mu]$. This ends the proof. □

Now, we prove that $KS^p[\mu]$ spaces are Banach lattices and Köthe function spaces. First, we recall some properties of such spaces (see also [43,44]).

A partially ordered Banach space Y, which is also a vector lattice, is a Banach lattice if $\|x\| \leq \|y\|$ for every $x, y \in Y$ with $|x| \leq |y|$.

A weak order unit of Y is a positive element $e \in Y$ such that, if $x \in Y$ and $x \wedge e = 0$, then $x = 0$.

Let Y be a Banach lattice and $\emptyset \neq A \subset B \subset Y$. We say that A is solid in B if for each x, y with $x \in B, y \in A$ and $|x| \leq |y|$, it is $x \in A$.

Let λ be an extended real-valued measure on Σ. A Banach space Y consisting of (classes of equivalence of) λ-measurable functions is called a Köthe function space with respect to λ if, for every $g \in Y$ and for each measurable function f with $|f| \leq |g|$ λ, almost everywhere, it is $f \in Y$ and $\|f\| \leq \|g\|$, and $\chi_A \in Y$ for every $A \in \Sigma$ with $\lambda(A) < +\infty$.

Theorem 4. *If $p \geq 1$, then $KS^p[\mu]$ is a Banach lattice with a weak order unit and a Köthe function space with respect to a control measure λ of μ.*

Proof. By the Rybakov theorem (see also [17] (Theorem IX.2.2)), there is $x_0' \in X'$ with $\|x_0'\| \leq 1$, such that $\lambda = x_0'\mu$ is a control measure of μ. If $f, g \in KS^p[\mu]$, $|f| \leq |g|\ \lambda$, almost everywhere, $k \in \mathbb{N}$ and $x' \in X'$ with $\|x'\| \leq 1$, then:

$$\left((L)\int_T \mathcal{E}_k(t)|f(t)|d|x'\mu|\right)^p \leq \left((L)\int_T \mathcal{E}_k(t)|g(t)|d|x'\mu|\right)^p \qquad (27)$$

(see also [16] (Proposition 5)), and hence, $\|f\|_{KS^p[\mu]} \leq \|g\|_{KS^p[\mu]}$. By (27), we can deduce that $KS^p[\mu]$ is a Banach lattice, because $KS^p[\mu]$ is the completion of $L^1[\mu]$ with respect to $\|\cdot\|_{KS^p[\mu]}$, $L^1[\mu]$ is a Banach lattice, and the lattice operations are continuous with respect to the norms (see also [44] (Proposition 1.1.6 (i))). Since $L^1[\mu]$ is solid with respect to the space of λ-measurable functions (see also [21]) and the closure of every solid subset of a Banach lattice is solid (see also [44] (Proposition 1.2.3 (i))), arguing similarly as in (27), we obtain that, if f is λ-measurable, $g \in KS^p[\mu]$, and $|f| \leq |g|\ \mu$, almost everywhere, then $g \in KS^p[\mu]$.

If $A \in \Sigma$, then $\lambda(A) < +\infty$ and $\chi_A \in L^1[\mu]$ (see also [16] (Proposition 5)), and hence, $\chi_A \in KS^p[\mu]$. Therefore, $KS^p[\mu]$ is a Köthe function space.

Finally, we prove that χ_T is a weak order unit of $KS^p[\mu]$. First, note that $\chi_T \in L^p[\mu]$, and hence, $\chi_T \in KS^p[\mu]$. Let $f \in KS^p[\mu]$ be such that $f^* = f \wedge \chi_T = 0\ \mu$, almost everywhere. We get:

$$\{t \in T : f^*(t) = 0\} = \{t \in T : f(t) = 0\},$$

and hence, $f = 0\ \mu$, almost everywhere. This ends the proof. □

Note that, by the definition of the (KL)-integral, the norm defined in (11) corresponds, in a certain sense, to the topology associated with the norm convergence of the integrals (μ-topology; see also [14] (Theorem 2.2.2)). However, with this norm, it is not natural to define an inner product in the space KS^2, since m is vector-valued.

On the other hand, when X' is separable and $\{x_h' : h \in \mathbb{N}\}$ is a countable dense subset of X', with $\|x_h'\| \leq 1$ for every h, it is possible to deal with the topology related to the weak convergence of integrals (weak μ-topology; see also [14] (Proposition 2.1.1)), whose corresponding norm is given by:

$$\|f\|_{KS^p[w\mu]} = \begin{cases} \left[\sum_{h=1}^{\infty} \omega_h \left(\sum_{k=1}^{\infty} \eta_k \left|(L)\int_T \mathcal{E}_k(t)f(t)\,d|x_h'\mu|\right|^p\right)\right]^{1/p} & \text{if } 1 \leq p < \infty, \\ \sup_{h\in\mathbb{N}}\left[\sup_{k\in\mathbb{N}}\left|(L)\int_T \mathcal{E}_k(t)f(t)d|x_h'\mu|\right|\right] & \text{if } p = \infty, \end{cases} \qquad (28)$$

where $\mathcal{E}_k$, $k \in \mathbb{N}$, is as in (11) and $(\eta_k)_k$, $(\omega_h)_h$ are two fixed sequences of strictly positive real numbers, such that $\sum_{k=1}^{\infty} \eta_k = \sum_{h=1}^{\infty} \omega_h = 1$. Note that, in general, a weak μ-topology does not coincide with a μ-topology, but there are some cases in which they are equal (see also [16] (Theorem 14)). Analogously, in Proposition 5, it is possible to prove the following:

Proposition 8. *For each $f \in L^1[\mu]$ and $p \geq 1$, it is:*

$$\|f\|_{KS^p[w\mu]} \leq \|f\|_{KS^\infty[w\mu]}. \qquad (29)$$

Now, we give the next fundamental result.

Theorem 5. *The map $f \mapsto \|f\|_{KS^p[w\mu]}$ defined in (28) is a norm.*

Proof. First of all, note that $\|f\|_{KS^p[\mu]} \geq 0$ for any $f \in L^1[\mu]$. Let $f \in L^1[\mu]$ be such that $\|f\|_{KS^p[\mu]} = 0$. We prove that $f = 0$ μ, almost everywhere. It will be enough to prove the assertion for $1 \leq p < \infty$, since the case $p = \infty$ follows from (29). Arguing analogously as in (15), we get:

$$\left|(L)\int_T \mathcal{E}_k(t)f(t)\,d|x'_h\mu|(t)\right| = 0 \quad \text{for every } h, k \in \mathbb{N}.$$

By contradiction, suppose that $f \neq 0$ μ, almost everywhere. If $E^+ = f^{-1}(]0, +\infty[)$, $E^- = f^{-1}(]-\infty, 0[)$, then $E^+, E^- \in \Sigma$, since f is Σ-measurable, and we have $\mu(E^+) \neq 0$ or $\mu(E^-) \neq 0$. Suppose that $\mu(E^+) \neq 0$. By the Hahn–Banach theorem, there is $x'_0 \in X'$ with $\|x'_0\| \leq 1$, $x'_0 \mu(E^+) \neq 0$, and hence, $|x'_0 \mu(E^+)| > 0$. Since the set $\{x'_h : h \in \mathbb{N}\}$ is dense in x' with respect to the norm of X', there is a positive integer h_0 with:

$$|x'_{h_0} \mu(E^+)| > 0. \tag{30}$$

Without loss of generality, we can assume $\|x'_{h_0}\| \leq 1$. Now, it is enough to proceed analogously as in Theorem 1, by replacing the linear continuous functional x'_0 in (18) with the element x'_{h_0} found in (30), by finding another element $x'_{h_1} \in X'$ with $|x'_{h_1}\mu(E^-)| > 0$, and by arguing again as in (18).

The triangular property of the norm is straightforward for $p = \infty$ and for $1 \leq p < \infty$ is a consequence of the inequality:

$$\left[\sum_{h=1}^\infty \omega_h \left(\sum_{k=1}^\infty \eta_k |b_{k,h} + c_{k,h}|^p\right)\right]^{1/p} \leq \left[\sum_{h=1}^\infty \omega_h \left(\sum_{k=1}^\infty \eta_k (|b_{k,h}| + |c_{k,h}|)^p\right)\right]^{1/p}$$

$$\leq \left[\sum_{h=1}^\infty \omega_h \left(\sum_{k=1}^\infty \eta_k |b_{k,h}|^p\right)\right]^{1/p} + \left[\sum_{h=1}^\infty \omega_h \left(\sum_{k=1}^\infty \eta_k |c_{k,h}|^p\right)\right]^{1/p} \tag{31}$$

which holds whenever $(b_{k,h})_{k,h}$, $(c_{k,h})_{k,h}$ are two double sequences of real numbers and $(\eta_k)_k$, $(\omega_h)_h$ are two sequences of positive real numbers, such that $\sum_{h=1}^\infty \omega_h = \sum_{k=1}^\infty \eta_k = 1$. The inequality in (31), as that in (10), follows from Minkowski's inequality. The other properties are easy to check. □

Now, in correspondence with the norm defined in (28), we define the following bilinear functional $\langle \cdot, \cdot \rangle : L^1[\mu] \times L^1[\mu] \to \mathbb{R}$ by:

$$\langle f, g \rangle_{KS^2[w\mu]} = \sum_{h=1}^\infty \omega_h \left[\sum_{k=1}^\infty \eta_k \left((L)\int_T \mathcal{E}_k(t)f(t)d|x'_h\mu|(t)\right)\left((L)\int_T \mathcal{E}_k(s)g(s)d|x'_h\mu|(s)\right)\right]. \tag{32}$$

Arguing similarly as in Theorem 5, it is possible to see that the functional $\langle \cdot, \cdot \rangle_{KS^2[w\mu]}$ in (32) is an inner product, and:

$$\|\cdot\|_{KS^2[w\mu]} = (\langle \cdot, \cdot \rangle_{KS^2[w\mu]})^{1/2}.$$

For $1 \leq p \leq \infty$, the Kuelbs–Steadman space $KS^p[w\mu]$ is the completion of $L^1[\mu]$ with respect to the norm defined in (28). Observe that, using Proposition 3, we can see that:

$$\|\cdot\|_{KS^p[w\mu]} \leq \|\cdot\|_{KS^p[\mu]} \text{ and } \|\cdot\|_{KS^p[w\mu]} \leq \|\cdot\|_{HKL} \text{ for } 1 \leq p \leq \infty.$$

As in Theorems 2 and 3, it is possible to prove the following:

Theorem 6. *For each p, q with $1 \leq p, q \leq \infty$, it is $L^q[\mu] \subset KS^p[w\mu]$ with continuous and dense embedding, and the space of all Σ-simple functions is dense in $KS^p[w\mu]$. Moreover, if $1 \leq p, q < \infty$, the embedding is completely continuous. Furthermore, $KS^p[w\mu]$ is a separable Banach lattice with a weak order unit and a Köthe function space with respect to a control measure λ of μ.*

Since $(KS^2[w\mu], \langle \cdot, \cdot \rangle_{KS^2[w\mu]})$ is a separable Hilbert space, by applying [2] (Theorems 5.15 and 8.7), it is possible to consider operators like, for instance, convolution and Fourier transform and to extend the theory there studied to the context of vector-valued measures (see also [45], [2] (Remark 5.16)).

4. Conclusions

We introduced Kuelbs–Steadman spaces related to the integration for scalar-valued functions with respect to a σ-additive measure μ, taking values in a Banach space X. We endowed them with the structure of the Banach space, both in connection with the norm convergence of integrals and in connection with the weak convergence of integrals ($KS^p[\mu]$ and $KS^p[w\mu]$, respectively). A fundamental role is played by the separability of μ. We proved that these spaces are separable Banach lattices and Köthe function spaces. Moreover, we saw that the embeddings of $L^q[\mu]$ into $KS^p[\mu]$ ($KS^p[w\mu]$) are continuous and dense, and also completely continuous when $1 \leq p, q < \infty$. When X' is separable, we endowed $KS^2[w\mu]$ with an inner product. In this case, $KS^2[w\mu]$ is a separable Hilbert space, and hence, it is possible to deal with operators like convolution and Fourier transform and to extend to Banach space-valued measures the theory investigated in [1–3].

Author Contributions: Conceptualization, B.H. and H.K.; methodology, A.B., B.H., and H.K.; validation, A.B.; formal analysis, A.B.; investigation, A.B., B.H., and H.K.; resources, A.B., B.H., and H.K.; data curation, A.B.; writing, original draft preparation, B.H. and H.K.; writing, review and editing, A.B.; visualization, A.B.; supervision, A.B.; project administration, A.B.; funding acquisition, A.B. All authors have read and agreed to the published version of the manuscript.

Funding: This research was partially supported by the project "Ricerca di Base 2019" (Metodi di approssimazione, misure, Analisi Funzionale, Statistica e applicazioni alla ricostruzione di immagini e documenti) of the University of Perugia and by the G.N.A.M.P.A.

Conflicts of Interest: The authors declare no conflict of interest.

References

1. Gill, T. L.; Myers, T. Constructive Analysis on Banach Spaces. *Real Anal. Exch.* **2019**, *44*, 1–36. [CrossRef]
2. Gill, T.L.; Zachary, W.W. *Functional Analysis and the Feynman Operator Calculus*; Springer: New York, NY, USA, 2016.
3. Kalita, H.; Hazarika, B.; Myers, T. Construction of the KS^p spaces on $\mathbb{R}^\infty$ and separable Banach spaces. *arXiv* **2020**, arXiv:2002.11512v2.
4. Kuelbs, J. Gaussian measures on a Banach space. *J. Funct. Anal.* **1970**, *5*, 354–367. [CrossRef]
5. Boccuto, A.; Candeloro, D.; Sambucini, A.R. Henstock multivalued integrability in Banach lattices with respect to pointwise non atomic measures. *Rend. Lincei Mat. Appl.* **2015**, *26*, 363–383. [CrossRef]
6. Candeloro, D.; Croitoru, A., Gavrilut, A.; Sambucini, A.R. An extension of the Birkhoff integrability for multifunctions. *Mediterr. J. Math.* **2016**, *13*, 2551–2575. [CrossRef]
7. Candeloro, D.; Di Piazza, L.; Musiał, K.; Sambucini, A.R. Gauge integrals and selections of weakly compact valued multifunctions. *J. Math. Anal. Appl.* **2016**, *441*, 293–308. [CrossRef]
8. Candeloro, D.; Di Piazza, L.; Musiał, K.; Sambucini, A.R. Some new results on integration for multifunctions. *Ric. Mat.* **2018**, *67*, 361–372. [CrossRef]
9. Candeloro, D.; Sambucini, A.R. Comparison between some norm and order gauge integrals in Banach lattices. *PanAm. Math. J.* **2015**, arXiv:1503.04968.
10. Caponetti, D.; Marraffa, V.; Naralenkov, K. On the integration of Riemann-measurable vector-valued functions. *Monatsh. Math.* **2017**, *182*, 513–536. [CrossRef]
11. Di Piazza, L.; Marraffa, V.; Musiał, K. Variational Henstock integrability of Banach space valued functions. *Math. Bohem.* **2016**, *141*, 287–296. [CrossRef]
12. Di Piazza, L.; Musiał, K. Set valued Kurzweil–Henstock–Pettis integral. *Set-Valued Anal.* **2005**, *13*, 167–169. [CrossRef]
13. Di Piazza, L.; Musiał, K. Decompositions of Weakly Compact Valued Integrable Multifunctions. *Mathematics* **2020**, *8*, 863. [CrossRef]

14. Ferrando Palomares, I. Duality in Spaces of *p*-Integrable Functions with Respect to a Vector Measure. Ph.D. Thesis, Universidad Politécnica de Valencia, Valencia, Spain, 2009.
15. Okada, S. The dual space of $\mathcal{L}^1(\mu)$ for a vector measure μ. *J. Math. Anal. Appl.* **1993**, *177*, 583–599. [CrossRef]
16. Sánchez-Pérez, E.A. Compactness arguments for spaces of *p*-integrable functions. *Ill. J. Math.* **2001**, *45*, 907–923.
17. Diestel, J.; Uhl, J.J. *Vector Measures*; American Mathematical Society: Providence, RI, USA, 1977.
18. Curbera, G.P.; Ricker, W.J. Vector Measures, Integration and Applications. In *Trends in Mathematics , Positivity*; Birkhäuser: Basel, Switzerland, 2007; pp. 127–160.
19. Kluvánek, I.; Knowles, G. *Vector Measures and Control Systems*; Mathematics Studies; North Holland Publ. Co.: Amsterdam, The Netherlands, 1975.
20. Lewis, D.R. Integration with respect to vector measures. *Pac. J. Math.* **1970**, *33*, 157–165. [CrossRef]
21. Panchapagesan, T.V. *The Bartle-Dunford-Schwartz Integral—Integration with Respect to a Sigma-Additive Vector Measure*; Birkhäuser: Basel, Switzerland; Boston, MA, USA; Berlin, Germany, 2008.
22. Ricker, W. Separability of the L^1-space of a vector measure. *Glasgow Math. J.* **1992**, *34*, 1–9. [CrossRef]
23. Brézis, H. *Functional Analysis, Sobolev Spaces and Partial Differential Equations*; Springer: New York, NY, USA; Dordrecht, The Netherlands; Heidelberg, Germany; London, UK, 2011.
24. Džamonja, M.; Kunen, K. Properties of the class of measure separable compact spaces. *Fund. Math.* **1995**, *147*, 261–277.
25. Boccuto, A.; Riečan, B.; Vrábelová, M. *Kurzweil–Henstock Integral in Riesz Spaces*; Bentham Sci. Publ.: Sharjah, UAE, 2009.
26. Candeloro, D.; Sambucini, A.R. Order-type Henstock and Mc Shane integrals in Banach lattice setting. In Proceedings of the SISY 2014—IEEE 12th International Symposium on Intelligent Systems and Informatics, Subotica, Serbia, 11–13 September 2014; pp. 55–59.
27. Cao, S. The Henstock integral for Banach-valued functions. *SEA Bull. Math.* **1992**, *16*, 35–40.
28. Federson, M. Some Peculiarities of the HK-integrals of Banach space-valued functions. *Real Anal. Exch.* **2004**, *29*, 439–460. [CrossRef]
29. Fremlin, D.H.; Mendoza, J. On the integration of vector-valued functions. *Ill. J. Math.* **1994**, *38*, 127–147. [CrossRef]
30. Gordon, R. *The Integrals of Lebesgue, Denjoy, Perron, and Henstock*; American Mathematical Society: Providence, RI, USA, 1994.
31. Lee, P.Y.; Výborný, R. *The Integral: An Easy Approach after Kurzweil and Henstock*; Cambridge University Press: Cambridge, UK, 2000.
32. Pfeffer, W.F. The multidimensional fundamental theorem of Calculus. *J. Austral. Math. Soc. (Ser. A)* **1987**, *43*, 143–170. [CrossRef]
33. Riečan, B. On the Kurzweil integral in compact topological spaces. *Radovi Mat.* **1986**, *2*, 151–163.
34. Curbera, G.P. Banach space properties of L^1 of a vector measure. *Proc. Am. Math. Soc.* **1995**, *123*, 3797–3806.
35. Fernández, A.; Mayoral, F.; Naranjo, F.; Sáez, C.; Sánchez-Pérez, E.A. Spaces of *p*-integrable functions with respect to a vector measure. *Positivity* **2006**, *10*, 1–6. [CrossRef]
36. Stefansson, G.F. L^1 of a vector measure. *Le Matematiche* **1993**, *48*, 219–234.
37. Boccuto, A.; Candeloro, D.; Sambucini, A.R. L^p-spaces in vector lattices and applications. *Math. Slovaca* **2017**, *67*, 1409–1426. [CrossRef]
38. Alexiewicz, A. Linear functionals on Denjoy integrable functions. *Colloq. Math.* **1948**, *1*, 289–293. [CrossRef]
39. Swartz, C. *Introduction to Gauge Integral*; World Scientific Publ.: Singapore, 2004.
40. McLeod, R.M. *The Generalized Riemann Integral*; Carus Math. Monograph 20; Mathematical Association of America: New York, NY, USA, 1980.
41. Hardy, G.H.; Littlewood, J.E.; Polya, G. *Inequalities*; Cambridge University Press: Cambridge, UK, 1934.
42. Kantorovich, L.A.; Akilov, G.P. *Functional Analysis*, 2nd ed.; Pergamon Press: Oxford, UK, 1982.
43. Lindenstrauss, J.; Tzafriri, L. *Classical Banach Spaces II, Function Spaces*; A Series of Modern Surveys in Mathematics; Springer: Berlin/Heidelberg, Germany; New York, NY, USA, 1979.

44. Meyer-Nieberg, P. *Banach Lattices*; Springer: Berlin/Heidelberg, Germany; New York, NY, USA, 1991.
45. Delgado, O.; Miana, P.J. Algebra structure for L^p of a vector measure. *J. Math. Anal. Appl.* **2009**, *358*, 355–363. [CrossRef]

© 2020 by the authors. Licensee MDPI, Basel, Switzerland. This article is an open access article distributed under the terms and conditions of the Creative Commons Attribution (CC BY) license (http://creativecommons.org/licenses/by/4.0/).

Article

Fractional Order of Evolution Inclusion Coupled with a Time and State Dependent Maximal Monotone Operator

Charles Castaing [1,*], Christiane Godet-Thobie [2] and Le Xuan Truong [3]

1. Département de Mathématiques, Université Montpellier II, Case Courrier 051, 34095 Montpellier CEDEX 5, France
2. Laboratoire de Mathématiques de Bretagne Atlantique, Université de Bretagne Occidentale, CNRS UMR 6205, 6, Avenue Victor Le Gorgeu, CS 9387, F-29238 Brest CEDEX 3, France; christiane.godet-thobie@univ-brest.fr
3. Department of Mathematics and Statistics, University of Economics Ho Chi Minh City, Ho Chi Minh City 700000, Vietnam; lxuantruong@gmail.com
* Correspondence: christiane.godet-thobie@univ-brest.fr

Received: 5 July 2020; Accepted: 11 August 2020; Published: 20 August 2020

Abstract: This paper is devoted to the study of evolution problems involving fractional flow and time and state dependent maximal monotone operator which is absolutely continuous in variation with respect to the Vladimirov's pseudo distance. In a first part, we solve a second order problem and give an application to sweeping process. In a second part, we study a class of fractional order problem driven by a time and state dependent maximal monotone operator with a Lipschitz perturbation in a separable Hilbert space. In the last part, we establish a Filippov theorem and a relaxation variant for fractional differential inclusion in a separable Banach space. In every part, some variants and applications are presented.

Keywords: fractional differential inclusion; maximal monotone operator; Riemann–Liouville integral; absolutely continuous in variation; Vladimirov pseudo-distance

MSC: 34H05; 34K35; 47H10; 28A25; 28B20; 28C20

1. Introduction

In recent decades, fractional equations and inclusions have proven to be interesting tools in the modeling of many physical or economic phenomena. In addition, there has been a significant development in fractional differential theory and applications in recent years [1–7]. In the case of the **sole inclusion**, $D^\alpha u(t) \subset \Gamma(t, u(t))$, one can find an important piece of literature. For examples, in following papers, study is made with different boundary conditions [8–12], with use of the non-compactness measure [13,14], with use of contraction principle in the space of selections of the set valued map instead in the space of solutions [15], with compactness conditions [16] or inclusions with infinite delay [17]. To the best of our acknowledge, a very few study is available in the fractional order differential inclusion **coupled** with a time and state dependent maximal monotone operator ([18] with subdifferential operators).

The main objective of the present work is to develop the existence theory for a **coupled system** of evolution inclusion driven by fractional differential equation and time and state dependent maximal monotone operators. The developments of the article are as follows.

At first, we investigate a second order problem governed a time and state dependent maximal monotone operator with Lipschitz perturbation in a separable Hilbert space E (The second order is in the state variable x).

$$(1.1) \begin{cases} x(t) = x_0 + \int_0^t u(s)ds, t \in [0,T] \\ u(t) \in D(A_{t,x(t)}), t \in [0,T] \\ -\dot{u}(t) \in A_{t,x(t)}u(t) + f(t,x(t),u(t)) \quad a.e. \end{cases}$$

Secondly, we investigate a class of fractional order problem driven by a time and state dependent maximal monotone operator with Lipschitz perturbation in E of the form

$$(1.2) \begin{cases} D^\alpha h(t) + \lambda D^{\alpha-1} h(t) = u(t), t \in [0,1] \\ I_{0^+}^\beta h(t) \vert_{t=0} := \lim_{t \to 0} \int_0^t \frac{(t-s)^{\beta-1}}{\Gamma(\beta)} h(s)ds = 0, \quad h(1) = I_{0^+}^\gamma h(1) = \int_0^1 \frac{(1-s)^{\gamma-1}}{\Gamma(\gamma)} h(s)ds \\ -\dot{u}(t) \in A_{t,h(t)}u(t) + f(t,h(t),u(t)) \quad a.e. \end{cases}$$

where $\alpha \in]1,2]$, $\beta \in [0, 2-\alpha]$, $\lambda \geq 0$, $\gamma > 0$ are given constants, D^α is the standard Riemann–Liouville fractional derivative, Γ is the gamma function, $(t,x) \to A_{(t,x)} : D(A_{(t,x)}) \to 2^E$ is a maximal monotone operator with domain $D(A_{(t,x)})$ and $f : [0,1] \times E \times E \to E$ is a single valued Lipschitz perturbation w.r.t $y \in E$.

Thirdly, we finish the paper with a Fillipov theorem and relaxation theorem for fractional differential inclusion in a separable Banach space E

$$(\mathcal{P}_F) \begin{cases} D^\alpha u(t) + \lambda D^{\alpha-1} u(t) \in F(t, u(t)), a.e. \ t \in [0,1] \\ I_{0^+}^\beta u(t) \vert_{t=0} = 0, \quad u(1) = I_{0^+}^\gamma u(1) \end{cases}$$

and

$$(\mathcal{P}_{\overline{co}F}) \begin{cases} D^\alpha u(t) + \lambda D^{\alpha-1} u(t) \in \overline{co}F(t, u(t)), a.e. \ t \in [0,1] \\ I_{0^+}^\beta u(t) \vert_{t=0} = 0, \quad u(1) = I_{0^+}^\gamma u(1) \end{cases}$$

where F is closed valued $\mathcal{L}(I) \times \mathcal{B}(E)$-measurable and Lipschitz w.r.t $x \in E$.

Within the framework of studies concerning coupled systems of evolution inclusion driven by fractional differential equation and time and state dependent maximal monotone operator, our results are fairly general and new and give further insight into the characteristics of both evolution inclusion and fractional order boundary value problems.

2. Notations and Preliminaries

In the whole paper, $I := [0, T]$ ($T > 0$) is an interval of $\mathbb{R}$ and E is a separable Hilbert space with the scalar product $\langle \cdot, \cdot \rangle$ and the associated norm $\| \cdot \|$. $\overline{B}_E$ denotes the unit closed ball of E and $r\overline{B}_E$ its closed ball of center 0 and radius $r > 0$. We denote by $\mathcal{L}(I)$ the sigma algebra on I, $\lambda := dt$ the Lebesgue measure and $\mathcal{B}(E)$ the Borel sigma algebra on E. If μ is a positive measure on I, we will denote by $L^p(I, E, \mu)$ $p \in [1, +\infty[$, (resp. $p = +\infty$), the Banach space of classes of measurable functions $u : I \to E$ such that $t \mapsto \|u(t)\|^p$ is μ-integrable (resp. u is μ-essentially bounded), equipped with its classical norm $\| \cdot \|_p$ (resp. $\| \cdot \|_\infty$). We denote by $\mathcal{C}(I, E)$ the Banach space of all continuous mappings $u : I \to E$, endowed with the sup norm.
The excess between closed subsets C_1 and C_2 of E is defined by $e(C_1, C_2) := \sup_{x \in C_1} d(x, C_2)$, and the Hausdorff distance between them is given by

$$d_H(C_1, C_2) := \max \left\{ e(C_1, C_2), e(C_2, C_1) \right\}.$$

The support function of $S \subset E$ is defined by: $\delta^*(a, S) := \sup_{x \in S} \langle a, x \rangle$, $\forall a \in E$. If X is a Banach space and X^* its topological dual, we denote by $\sigma(X, X^*)$ the weak topology on X, and by $\sigma(X^*, X)$ the weak* topology on X^*.

Let $A : E \rightrightarrows E$ be a set-valued map. We denote by $D(A)$, $R(A)$ and $Gr(A)$ its domain, range and graph. We say that A is monotone, if $\langle y_1 - y_2, x_1 - x_2 \rangle \geq 0$ whenever $x_i \in D(A)$, and $y_i \in A(x_i)$, $i = 1, 2$. In addition, we say that A is a maximal monotone operator of E, if its graph could not be contained properly in the graph of any other monotone operator. By Minty's Theorem, A is maximal monotone iff $R(I_E + A) = E$.

If A is a maximal monotone operator of E, then, for every $x \in D(A)$, $A(x)$ is nonempty closed and convex. We denote the projection of the origin on the set $A(x)$ by $A^0(x)$.

Let $\lambda > 0$; then, the resolvent and the Yosida approximation of A are the well-known operators defined respectively by $J_\lambda^A = (I_E + \lambda A)^{-1}$ and $A_\lambda = \frac{1}{\lambda}(I_E - J_\lambda^A)$. These operators are single-valued and defined on all of E, and we have $J_\lambda^A(x) \in D(A)$, for all $x \in E$. For more details about the theory of maximal monotone operators, we refer the reader to [5,19,20].

Let $A : D(A) \subset E \to 2^E$ and $B : D(B) \subset E \to 2^E$ be two maximal monotone operators, then we denote by $dis(A, B)$ the pseudo-distance between A and B defined by

$$dis(A, B) = \sup\left\{ \frac{\langle y - y', x' - x \rangle}{1 + \|y\| + \|y'\|} : x \in D(A), y \in Ax, x' \in D(B), y' \in Bx' \right\}. \tag{1}$$

This pseudo-distance due to Vladimiro [21] is particularly well suited to the study of operators (see its use in [22]) and also, in the sweeping process, for its links with the Hausdorff distance in convex analysis. Indeed, if $N_{C(t,x)}$ is the normal cone of the closed convex set $C(t, x)$, we have

$$dis(N_{C(t,x)}, N_{C(s,y)}) = d_H(C(t, x), C(s, y)).$$

This property will be used in this paper.

For the proof of our main theorems, we will need some elementary lemmas taken from reference [23].

Lemma 1. *Let A be a maximal monotone operator of E. If $x \in \overline{D(A)}$ and $y \in E$ are such that*

$$\langle A^0(z) - y, z - x \rangle \geq 0 \ \forall z \in D(A),$$

then $x \in D(A)$ and $y \in A(x)$.

Lemma 2. *Let A_n ($n \in \mathbb{N}$), A be maximal monotone operators of E such that $dis(A_n, A) \to 0$. Suppose also that $x_n \in D(A_n)$ with $x_n \to x$ and $y_n \in A_n(x_n)$ with $y_n \to y$ weakly for some $x, y \in E$. Then, $x \in D(A)$ and $y \in A(x)$.*

Lemma 3. *Let A, B be maximal monotone operators of E. Then,*
(1) for $\lambda > 0$ and $x \in D(A)$

$$\|x - J_\lambda^B(x)\| \leq \lambda \|A^0(x)\| + dis(A, B) + \sqrt{\lambda(1 + \|A^0(x)\|)dis(A, B)}.$$

(2) For $\lambda > 0$ and $x, x' \in E$

$$\|J_\lambda^A(x) - J_\lambda^A(x')\| \leq \|x - x'\|.$$

Lemma 4. *Let A_n ($n \in \mathbb{N}$), A be maximal monotone operators of E such that $dis(A_n, A) \to 0$ and $\|A_n^0(x)\| \leq c(1 + \|x\|)$ for some $c > 0$, all $n \in \mathbb{N}$ and $x \in D(A_n)$. Then, for every $z \in D(A)$, there exists a sequence (ζ_n) such that*

$$\zeta_n \in D(A_n), \quad \zeta_n \to z \ \text{and} \ A_n^0(\zeta_n) \to A^0(z). \tag{2}$$

3. On Second Order Problem Driven by a Time and State Dependent Maximal Operator

Let $I = [0, T]$ and let E be a separable Hilbert space. In this part, we are interested in solving the problem (1.1).

Lemma 5. *Let $(t, x) \to A_{(t,x)} : D(A_{(t,x)}) \to 2^E$ a maximal monotone operator satisfying:*
(H_1) $\|A^0_{(t,x)} y\| \leq c(1 + \|x\| + \|y\|)$ *for all* $(t, x, y) \in I \times E \times D(A_{(t,x)})$, *for some positive constant c,*
(H_2) $dis(A_{(t,x)}, A_{(\tau,y)}) \leq a(t) - a(\tau) + r\|x - y\|$, *for all $0 \leq \tau \leq t \leq T$ and for all $(x, y) \in E \times E$, where r is a positive number, $a : I \to [0, +\infty[$ is nondecreasing absolutely continuous on I with $\dot{a} \in L^2$, shortly $a \in W^{1,2}(I)$.*
Then, the following hold:

Fact $\mathcal{I}$: *For any absolutely continuous $x \in W^{1,2}_E(I)$ and for any $u_0 \in D(A_{(0,x(0))})$, the problem*

$$\begin{cases} -\dot{u}(t) \in A_{(t,x(t))} u(t), \text{ a.e. } t \in I \\ u(t) \in D(A_{(t,x(t))}), \forall t \in I \\ u(0) = u_0 \in D(A_{(0,x(0))}) \end{cases}$$

has a unique absolutely continuous solution with $\|\dot{u}(t)\| \leq K(1 + \dot{\beta}(t))$ where $\beta(t) = \int_0^t [\dot{a}(s) + r\|\dot{x}(s)\|] ds$, $\forall t \in I$ and K is a positive constant depending on $\|u_0\|, c, T, x$ and β.

Fact $\mathcal{J}$: *Assume that*
(H_3) $(t, x, y) \to J^{A_{(t,x)}}_\lambda (y)$ *is $\mathcal{L}(I) \otimes \mathcal{B}(E) \otimes \mathcal{B}(E)$-measurable.*
Then, the composition operator $\mathcal{A}_x : D(\mathcal{A}_x) \subset L^2(I, E, dt) \to 2^{L^2(I,E,dt)}$ defined by

$$\mathcal{A}_x u = \{v \in L^2(I, E, dt) : v(t) \in A_{(t,x(t))} u(t) \text{ a.e. } t \in I\}$$

for each $u \in D(\mathcal{A}_x)$ where
$D(\mathcal{A}_x) := \{u \in L^2(I, E, dt) : u(t) \in D(A_{(t,x(t))}) \text{ a.e. } t \in I, \text{ for which } \exists y \in L^2(I, E, dt) : y(t) \in A_{(t,x(t))} u(t), \text{ a.e. } t \in I\}$
is maximal monotone. Consequently, the graph of $\mathcal{A}_x : D(\mathcal{A}_x) \subset L^2(I, E, dt) \to 2^{L^2(I,E,dt)}$ is strongly-weakly sequentially closed in $L^2(I, E, dt) \times L^2(I, E, dt)$.

Proof. Fact $\mathcal{I}$. The mapping $B_t = A_{(t,x(t))}$ is a time dependent absolutely continuous in variation maximal monotone operator: For all $0 \leq \tau \leq t \leq T$, we have by (H_2)

$$\begin{cases} dis(B_t, B_\tau) = dis(A_{(t,x(t))}, A_{(\tau,x(\tau))}) \\ \leq |a(t) - a(\tau)| + r\|x(t) - x(\tau)\| \\ \leq \int_\tau^t \dot{a}(s) ds + r \int_\tau^t \|\dot{x}(s)\| ds \\ = \beta(t) - \beta(\tau) \end{cases}$$

where $\beta(t) = \int_0^t [\dot{a}(s) + r\|\dot{x}(s)\|] ds$, $\forall t \in I$. Furthermore, by (H_1), we have

$$\begin{cases} \|B_t^0 y\| = \|A^0_{(t,x(t))} y\| \leq c(1 + \|x(t)\| + \|y\|) \\ \leq c_1(1 + \|y\|) \end{cases}$$

for all $y \in D(A_{(t,x(t))})$, where c_1 is a positive generic constant. Consequently, by [22] (Theorem 3.5), for every $u_0 \in D(B_0)$, a unique absolutely continuous mapping $u : I \to E$ exists satisfying

$$\begin{cases} -\dot{u}(t) \in B_t u(t) = A_{(t,x(t))} u(t), \text{ a.e. } t \in I \\ u(t) \in D(B_t) = D(A_{(t,x(t))}), \forall t \in I \\ u(0) = u_0 \in D(B_0) = D(A_{(0,x(0))}) \end{cases}$$

with $\|\dot{u}(t)\| \leq K(1 + \dot{\beta}(t))$, where $\beta(t) = \int_0^t [\dot{a}(s) + r\|\dot{x}(s)\|]ds$, $\forall t \in I$ and K is a positive constant depending on $\|u_0\|, c, T, \beta$.

Fact $\mathcal{J}$. Taking account $\mathcal{J}$, it is clear that $D(\mathcal{A}_x)$ is nonempty and $\mathcal{A}_x$ is well defined. It is easy to see that $\mathcal{A}_x$ is monotone. Let us prove that $\mathcal{A}_x$ is maximal monotone. We have to check that $R(I_{L^2(I,E,dt)} + \lambda \mathcal{A}_x) = L^2(I, E, dt)$ for each $\lambda > 0$. Let $g \in L^2(I, E, dt)$. Then, from (H3) $t \mapsto v(t) = J_\lambda^{A_{(t,x(t))}} g(t) = g(t) - \lambda A_\lambda^{A_{(t,x(t))}} g(t)$ is measurable. Set

$$h(t) = \lambda A_\lambda^{A_{(t,x(t))}} g(t) = \lambda A_\lambda^{A_{(t,x(t))}} g(t) - \lambda A_\lambda^{A_{(t,x(t))}} u(t) + \lambda A_\lambda^{A_{(t,x(t))}} u(t)$$

where u denotes the absolutely continuous solution to $-\frac{du}{dt}(t) \in A_{(t,x(t))} u(t)$ using **Fact $\mathcal{I}$**. Then, h is measurable with

$$\|h(t)\| \leq 2\|g(t) - u(t)\| + \lambda \|A_\lambda^{A_{(t,x(t))}} u(t)\|$$

by noting that $A_\lambda^{A_{(t,x(t))}}$ is $\frac{2}{\lambda}$-Lipschitz and so we deduce that $h \in L^2(I, E, dt)$ because $g \in L^2(I, E, dt)$ and $t \mapsto A_\lambda^{A_{(t,x(t))}} u(t) \in L^\infty(I, E, dt)$ using (H1). This proves that $v \in L^2(I, E, dt)$ and $g \in v + \lambda \mathcal{A}_x v$ so that $R(I_{L^2(I,E;dt)} + \lambda \mathcal{A}_x) = L^2(I, E, dt)$. $\square$

Here is a useful application.

Corollary 1. *With hypotheses and notation of the preceding lemma, let (v_n) and (u_n) be two sequences in $L^2(I, E, dt)$ such that $v_n(t) \in A_{(t,x(t))} u_n(t)$ a.e for all $n \in \mathbf{N}$. If $v_n \to v$ weakly in $L^2(I, E, dt)$ and $u_n \to u$ strongly in $L^2(I, E, dt)$, then $v(t) \in A_{(t,x(t))} u(t)$ a.e.*

Theorem 1. *Let $I = [0, T]$. Let $(t, x) \to A_{(t,x)} : D(A_{(t,x)}) \to 2^E$ a maximal monotone operator satisfying:*
(H$_1$) $\|A_{(t,x)}^0 y\| \leq c(1 + \|x\| + \|y\|)$ for all $(t, x, y) \in I \times E \times D(A_{(t,x)})$, for some positive constant c,
(H$_2$) $dis(A_{(t,x)}, A_{(\tau,y)}) \leq a(t) - a(\tau) + r\|x - y\|$, for all $0 \leq \tau \leq t \leq T$ and for all $(x, y) \in E \times E$, where r is a positive number, $a : I \to [0, +\infty[$ is nondecreasing absolutely continuous on I with $\dot{a} \in L^2(I, \mathbf{R}, dt)$,
(H$_3$) $D(A_{(t,x)})$ is boundedly-compactly measurable in the sense, for any bounded set $B \subset E$, there is a measurable compact valued integrably bounded mapping $\Psi_B : I \to E$ such that $D(A_{(t,x)}) \subset \Psi_B(t) \subset \gamma(t)\overline{B}_E$ for all $(t, x) \in I \times B$ where $\gamma \in L^2(I, \mathbf{R}, dt)$.

Then, for any $(x_0, u_0) \in E \times D(A_{(0,x_0)})$, there exist an absolutely continuous $x : I \to E$ and an absolutely continuous $u : I \to E$ such that

$$\begin{cases} x(t) = x_0 + \int_0^t u(s)ds, & \forall t \in I \\ x(0) = x_0, u(0) = u_0 \in D(A_{(0,x_0)}) \\ -\dot{u}(t) \in A_{(t,x(t))} u(t) & \text{a.e. } t \in I \\ u(t) \in D(A_{(t,x(t))}), \forall t \in I \end{cases}$$

Proof. Let us consider the closed convex subset $\mathcal{X}_\gamma$ in the Banach space $\mathcal{C}_E(I)$ defined by

$$\mathcal{X}_\gamma : \{h \in W^{1,2}(I, E) : h(t) = x_0 + \int_0^t \dot{h}(s)ds, \|\dot{h}(s)\| \leq \gamma(s) \text{ a.e.}, \gamma \in L^2(I, \mathbf{R}, dt)\}.$$

Then, $\mathcal{X}_\gamma$ is equi-absolutely continuous. By the fact that $\mathcal{J}$, for each $h \in \mathcal{X}_\gamma$, there is a unique $W^{1,2}(I, E)$ mapping $u_h : I \to E$, which is the $W^{1,2}(I, E)$ solution to the inclusion

$$\begin{cases} -\dot{u}_h(t) \in A_{(t,h(t))} u_h(t) & \text{a.e. } t \in I \\ u_h(t) \in D(A_{(t,h(t))}), \forall t \in I \\ u_h(0) = u_0 \in D(A_{(0,h(0))}) = D(A_{(0,x_0)}), \end{cases}$$

with $\|\dot{u}_h(t)\| \leq K(1 + \dot{\beta}(t))$, where $\beta(t) = \int_0^t [\dot{a}(s) + \gamma(s)] ds$, $\forall t \in I$ and K is a positive constant depending on $\|u_0\|, c, T, \beta$. We refer to [22] (Theorem 3.5) for details of the estimate of the velocity. Now, for each $h \in \mathcal{X}_\gamma$, let us consider the mapping

$$\Phi(h)(t) := x_0 + \int_0^t u_h(s) ds, \ t \in I.$$

As $u_h(s) \in D(A_{(s,h(s))}) \subset \bigcup_{x \in \mathcal{X}_\gamma(s)} D(A_{(s,x)}) \subset \Psi_\gamma(s) \subset \gamma(s)\overline{B}_E$ for all $s \in [0, T]$, where $\Psi_\gamma : I \to E$ is a compact valued measurable mapping given by condition (H_3). It is clear that $\Phi(h) \in \mathcal{X}_\gamma$. Our aim is to prove the existence theorem by applying some ideas developed in [24] via a generalized fixed point theorem [25] (Theorem 4.3), [26] (Lemma 1). Nevertheless, this needs a careful look using the estimation of the absolutely continuous solution given above. For this purpose, we first claim that $\Phi : \mathcal{X}_\gamma \to \mathcal{X}_\gamma$ is continuous and, for any $h \in \mathcal{X}_\gamma$ and for any $t \in I$, the inclusion holds

$$\Phi(h)(t) \in u_0 + \int_0^t \overline{co}\Psi_\gamma(s) ds.$$

Since $s \mapsto \overline{co}\Psi_\gamma(s)$ is a convex compact valued and integrably bounded multifunction, the second member is convex compact valued [27] so that $\Phi(\mathcal{X})$ is equicontinuous and relatively compact in the Banach space $\mathcal{C}_E(I)$. Now, we check that Φ is continuous. It is sufficient to show that, if (h_n) converges uniformly to h in $\mathcal{X}_\gamma$, then the AC solution u_{h_n} associated with h_n

$$\begin{cases} u_{h_n}(0) \in D(A_{(0,h_n(0))}) \\ u_{h_n}(t) \in D(A_{(t,h_n(t))}), \forall t \in I \\ -\dot{u}_{h_n}(t) \in A_{(t,h_n(t))} u_{h_n}(t) & \text{a.e. } t \in I \end{cases}$$

uniformly converges to the AC solution u_h associated with h

$$\begin{cases} u_h(0) = u_0 \in D(A_{(0,h(0))}) \\ u_h(t) \in D(A_{(t,h(t))}), \forall t \in I \\ -\dot{u}_h(t) \in A_{(t,h(t))} u_h(t) & \text{a.e. } t \in I \end{cases}$$

As (u_{h_n}) is equi-absolutely continuous with the estimate $\|\dot{u}_{h_n}(t)\| \leq K(1 + \dot{\beta}(t))$ a.e for all $n \in \mathbf{N}$, we may assume that (u_{h_n}) converges uniformly to a AC mapping u and $(\frac{du_{h_n}}{dt})$ converges weakly in $L_E^2(I, dt)$ to $w \in L_E^2(I, dt)$ with $\|w(t)\| \leq K(1 + \dot{\beta}(t))$ a.e. $t \in I$ so that

$$\text{weak-}\lim_n u_{h_n} = \text{weak-}\lim_n u_{h_n}(0) + \text{weak-}\lim_n \int_I \frac{du_{h_n}}{dt}$$

$$= u(0) + \int_I w \, dt := z(t), \ t \in I$$

By identifying the limits, we get $u(t) = z(t) = u(0) + \int_I w\, dt$, $t \in I$ with $u(0) =$ weak-$\lim_n u_{h_n}(0) = \lim_n u_{h_n}(0)$ and $\frac{du}{dt} = w$. As $u_{h_n}(t) \in D(A_{(t,h_n(t))})$, $\forall t \in I$ and $u_{h_n}(t) \to u(t)$, $A^0_{(t,h_n(t))} u_{h_n}(t)$ is bounded using (H_1) for every $t \in [0,T]$ and

$$dis(A_{(t,h_n(t))}, A_{(t,h(t))}) \leq r\|h_n(t) - h(t)\| \to 0$$

when $n \to \infty$ by (H_2), from Lemma 2, we deduce that $u(t) \in D(A_{(t,h(t))})$, $\forall t \in I$. Now, we are going to check that u satisfies the inclusion

$$-\frac{du}{dt}(t) \in A_{(t,h(t))} u(t) \quad \text{a.e. } t \in I$$

As $\frac{du_{h_n}}{dt} \to \frac{du}{dt}$ weakly in $L^2(I, E, dt)$, we may assume that $(\frac{du_{h_n}}{dt})$ Komlos converges to $\frac{du}{dt}$. There is a dt-negligible set N such that for $t \in I \setminus N$

$$\lim_{n \to \infty} \frac{1}{n} \sum_{j=1}^n \frac{du_{h_j}}{dt}(t) = \frac{du}{dt}(t). \tag{3}$$

$$-\frac{du_{h_n}}{dt}(t) \in A_{(t,h_n(t))} u_n(t). \tag{4}$$

Let $\eta \in D(A_{(t,h(t))})$.

Using Lemma 4, there is a sequence (η_n) such that $\eta_n \in D(A_{(t,h_n(t))})$, $\eta_n \to \eta$ and $A^0_{(t,h_n(t))} \eta_n \to A^0_{(t,h(t))} \eta$. From (4), by monotonicity,

$$\langle \frac{du_{h_n}}{dt}, u_{h_n}(t) - \eta_n \rangle \leq \langle A^0_{(t,h_n(t))} \eta_n, \eta_n - u_{h_n}(t) \rangle. \tag{5}$$

From

$$\langle \frac{du_{h_n}}{dt}(t), u(t) - \eta \rangle = \langle \frac{du_{h_n}}{dt}(t), u_{h_n}(t) - \eta_n \rangle + \langle \frac{du_{h_n}}{dt}(t), u(t) - u_{h_n}(t) - (\eta - \eta_n) \rangle,$$

let us write

$$\frac{1}{n} \sum_{j=1}^n \langle \frac{du_{h_j}}{dt}(t), u(t) - \eta \rangle = \frac{1}{n} \sum_{j=1}^n \langle \frac{du_{h_j}}{dt}(t), u_{h_j}(t) - \eta_j \rangle + \frac{1}{n} \sum_{j=1}^n \langle \frac{du_{h_j}}{dt}(t), u(t) - u_{h_j}(t) \rangle$$

$$+ \sum_{j=1}^n \langle \frac{du_{h_j}}{dt}(t), \eta_j - \eta \rangle,$$

so that

$$\frac{1}{n} \sum_{j=1}^n \langle \frac{du_{h_j}}{dt}(t), u(t) - \eta \rangle \leq \frac{1}{n} \sum_{j=1}^n \langle A^0_{(t,h_j(t))} \eta_j, \eta_j - u_{h_j}(t) \rangle + K(1 + \dot{\beta}(t)) \frac{1}{n} \sum_{j=1}^n \|u(t) - u_{h_j}(t)\|.$$

$$+ K(1 + \dot{\beta}(t)) \frac{1}{n} \sum_{j=1}^n \|\eta_j - \eta\|.$$

Passing to the limit using (3) when $n \to \infty$, this last inequality gives immediately

$$\langle \frac{du}{dt}(t), u(t) - \eta \rangle \leq \langle A^0_{(t,h(t))} \eta, \eta - u(t) \rangle \quad \text{a.e.}$$

As a consequence, by Lemma 1, we get $-\frac{du}{dt}(t) \in A_{(t,h(t))}u(t)$ a.e. with $u(0) \in D(A_{(0,h(0))})$ so that, by uniqueness, $u = u_h$.

Now, let us check that $\Phi : \mathcal{X} \to \mathcal{X}$ is continuous. Let $h_n \to h$. We have

$$\Phi(h_n)(t) - \Phi(h)(t) = \int_0^t u_{h_n}(s)ds - \int_0^t u_h(s)ds = \int_0^t [u_{h_n}(s) - u_h(s)]ds$$

As $||u_{h_n}(.) - u_h(.)|| \to 0$ pointwisely and is uniformly bounded, we conclude that

$$\sup_{t \in I} ||\Phi(h_n)(t) - \Phi(h)(t)|| \leq \sup_{t \in I} \int_0^t ||u_{h_n}(.) - u_h(.)||ds \to 0$$

so that $\Phi(h_n) - \Phi(h) \to 0$ in $\mathcal{C}_E(I)$. Since $\Phi : \mathcal{X}_\gamma \to \mathcal{X}_\gamma$ is continuous and $\Phi(\mathcal{X}_\gamma)$ is relatively compact in $\mathcal{C}_E(I)$, by [25] (Theorem 4.3), [26] (Lemma 1), Φ has a fixed point, say $h = \Phi(h) \in \mathcal{X}_\gamma$ that means

$$h(t) = \Phi(h)(t) = x_0 + \int_0^t u_h(s)ds, \ t \in I,$$

$$\begin{cases} u_h(t) \in D(A_{(t,h(t))}) \\ -\dfrac{du_h}{dt}(t) \in A_{(t,h(t))}u_h(t) \ dt\text{-a.e.} \end{cases}$$

the proof is complete. □

There is a direct application to sweeping process.

Corollary 2. *Let $C : I \times E \to E$ be a convex compact valued mapping satisfying*
(i) $C(t,x) \subset \gamma(t)\overline{B}_E, \forall (t,x) \in I \times E$, where $\gamma \in L^2(I, \mathbb{R}, dt)$,
(ii) $d_H(C(s,x), C(t,y)) \leq a(t) - a(\tau) + r||x-y||$, for all $0 \leq \tau \leq t \leq 1$ and for all $(x,y) \in E \times E$, where r is a positive number, $a : I \to [0, +\infty[$ is nondecreasing absolutely continuous on I with $\dot{a} \in L^2(I, \mathbb{R}, dt)$,
(iii) For any $t \in I$, for any bounded set $B \subset E$, $C(t, B)$ is relatively compact.
Then, for any $(x_0, u_0) \in E \times C(0, x_0)$, there exist an absolutely continuous $x : I \to E$ and and absolutely continuous $u : I \to E$ such that

$$\begin{cases} x(t) = x_0 + \int_0^t u(s)ds, \quad \forall t \in I \\ x(0) = x_0, u(0) = u_0 \in C(0, x_0) \\ -\dot{u}(t) \in N_{C(t,x(t))}u(t) \quad \text{a.e. } t \in I \\ u(t) \in C(t, x(t)), \forall t \in I \end{cases}$$

Proof. It is easy to apply Theorem 1 with $A_{(t,x(t))} = N_{C(t,x(t))}$ □

Now, we proceed to the Lipschitz perturbation of the preceding theorem.

Theorem 2. *Let $I = [0, T]$. Let $(t,x) \to A_{(t,x)} : D(A_{(t,x)}) \to 2^E$ be a maximal monotone operator satisfying:*
(H_1) $||A^0_{(t,x)}y|| \leq c(1 + ||x|| + ||y||)$ for all $(t,x,y) \in I \times E \times D(A_{(t,x)})$, for some positive constant c,
(H_2) $dis(A_{(t,x)}, A_{(\tau,y)}) \leq a(t) - a(\tau) + r||x-y||$, for all $0 \leq \tau \leq t \leq T$ and for all $(x,y) \in E \times E$, where r is a positive number, $a : I \to [0, +\infty[$ is nondecreasing absolutely continuous on I with $\dot{a} \in L^2(I, \mathbb{R}, dt)$,
(H_3) $D(A_{(t,x)})$ is boundedly-compactly measurable in the sense, for any bounded set $B \subset E$, there is a measurable compact valued integrably bounded mapping $\Psi_B : I \to E$ such that $D(A_{(t,x)}) \subset \Psi_B(t) \subset \gamma(t)\overline{B}_E$ for all $(t,x) \in I \times B$, where $\gamma \in L^2(I, \mathbb{R}, dt)$.
Let $f : I \times E \times E \to E$ such that
(i) $f(., x, y)$ is Lebesgue measurable on I for all $(x, y) \in E \times E$
(ii) $f(t, ., .)$ is continuous on $E \times E$,
(iii) $||f(t, x, y)|| \leq M$ for all $(t, x, y) \in I \times E \times E$,
(iv) $||f(t, x, y) - f(t, x, z)|| \leq M||y - z||$, for all $(t, x, y, z) \in I \times E \times E \times E$

for some positive constant M.
Then, for any $(x_0, u_0) \in E \times D(A_{(0,x_0)})$, there exists an absolutely continuous $x : I \to E$ and an absolutely continuous $u : I \to E$ such that

$$\begin{cases} x(t) = x_0 + \int_0^t u(s)ds, & \forall t \in I \\ x(0) = x_0, u(0) = u_0 \in D(A_{(0,x_0)}) \\ -\dot{u}(t) \in A_{(t,x(t))} u(t) + f(t, x(t), u(t)) & \text{a.e. } t \in I \\ u(t) \in D(A_{(t,x(t))}), \forall t \in I \end{cases}$$

Proof. Let us consider the closed convex subset $\mathcal{X}_\gamma$ in the Banach space $\mathcal{C}_E(I)$ defined by

$$\mathcal{X}_\gamma : \{h \in W^{1,2}(I, E) : h(t) = x_0 + \int_0^t \dot{h}(s)ds, ||\dot{h}(s)|| \leq \gamma(s) \text{ a.e.}, \gamma \in L^2(I, \mathbf{R}, dt)\}.$$

Then, $\mathcal{X}_\gamma$ is equi-absolutely continuous. By fact $\mathcal{J}$, for each $h \in \mathcal{X}_\gamma$, there is a unique $W^{1,2}(I, E)$ mapping $u_h : I \to E$, which is the $W^{1,2}(I, E)$ solution to the inclusion

$$\begin{cases} -\dot{u}_h(t) \in A_{(t,h(t))} u_h(t) + f(t, h(t), u_h(t)) & \text{a.e. } t \in I \\ u_h(t) \in D(A_{(t,h(t))}), \forall t \in I \\ u_h(0) = u_0 \in D(A_{(0,h(0))}) = D(A_{(0,x_0)})), \end{cases}$$

with $||\dot{u}_h(t)|| \leq K(1 + \dot{\beta}(t)) + M(K + 1) = \eta(t)$ where $\beta(t) = \int_0^t [\dot{a}(s) + \gamma(s)]ds, \forall t \in I$ and K is a positive constant depending on $||u_0||, c, T, \beta$. We refer to (Theorem 3.5) for details of the estimate of the velocity. Now, for each $h \in \mathcal{X}_\gamma$, let us consider the mapping

$$\Phi(h)(t) := x_0 + \int_0^t u_h(s)ds, \ t \in I.$$

As $u_h(s) \in D(A_{(s,h(s))}) \subset \bigcup_{x \in \mathcal{X}_\gamma(s)} D(A_{(s,x)}) \subset \Psi_\gamma(s) \subset \gamma(s)\overline{B}_E$ for all $s \in [0, T]$, where $\Psi_\gamma : I \to E$ is a compact valued measurable mapping given by condition (H_3). It is clear that $\Phi(h) \in \mathcal{X}_\gamma$. Our aim is to prove the existence theorem by applying some ideas developed in Castaing et al. [24] via the same generalized fixed point theorem already used [25,26]. Nevertheless, this needs a careful look using the estimation of the absolutely continuous solution given above. For this purpose, we first claim that $\Phi : \mathcal{X}_\gamma \to \mathcal{X}_\gamma$ is continuous, and, for any $h \in \mathcal{X}_\gamma$ and for any $t \in I$, the inclusion holds

$$\Phi(h)(t) \in u_0 + \int_0^t \overline{co}\Psi_\gamma(s)ds.$$

Since $s \mapsto \overline{co}\Psi_\gamma(s)$ is a convex compact valued and integrably bounded multifunction, the second member is convex compact valued [27] so that $\Phi(\mathcal{X})$ is equicontinuous and relatively compact in the Banach space $\mathcal{C}_E(I)$. Now, we check that Φ is continuous. It is sufficient to show that, if (h_n) converges uniformly to h in $\mathcal{X}_\gamma$, then the AC solution u_{h_n} associated with h_n

$$\begin{cases} u_{h_n}(0) \in D(A_{(0,h_n(0))}) \\ u_{h_n}(t) \in D(A_{(t,h_n(t))}), \forall t \in I \\ -\dot{u}_{h_n}(t) \in A_{(t,h_n(t))} u_{h_n}(t) + f(t, h_n(t), u_{h_n}(t)), & \text{a.e. } t \in I \end{cases}$$

uniformly converges to the AC solution u_h associated with h

$$\begin{cases} u_h(0) \in D(A_{(0,h(0))}) \\ u_h(t) \in D(A_{(t,h(t))}), \forall t \in I \\ -\dot{u}_h(t) \in A_{(t,h(t))} u_h(t) + f(t, h(t), u_h(t)) & \text{a.e. } t \in I \end{cases}$$

As (u_{h_n}) is equi-absolutely continuous with the estimate $||\dot{u}_{h_n}(t)|| \leq K(1+\dot{\beta}(t)) + (K+1)M = \psi(t)$ a.e for all $n \in \mathbf{N}$, we may assume that (u_{h_n}) converges uniformly to a AC mapping u and $(\frac{du_{h_n}}{dt})$ converges weakly in $L^2_E(I, dt)$ to $w \in L^2_E(I, dt)$ with $||w(t)|| \leq K(1+\dot{\beta}(t)) + (K+1)M$ a.e. $t \in I$ so that

$$\text{weak-}\lim_n u_{h_n} = \text{weak-}\lim_n u_{h_n}(0) + \text{weak-}\lim_n \int_{[0,t]} \frac{du_{h_n}}{dt}$$

$$= u(0) + \int_{[0,t]} w \, dt := z(t), \ t \in I$$

By identifying the limits, we get
$u(t) = z(t) = u(0) + \int_{[0,t]} w \, dt, \ t \in I$ with $u(0) = \text{weak-}\lim_n u_{h_n}(0) = \lim_n u_{h_n}(0)$ and $\frac{du}{dt} = w$. As $u_{h_n}(t) \in D(A_{(t,h_n(t))}), \forall t \in I$ and $u_{h_n}(t) \to u(t)$, $A^0_{(t,h_n(t))} u_{h_n}(t)$ is bounded using (H_1) for every $t \in I$ and

$$dis(A_{(t,h_n(t))}, A_{(t,h(t))}) \leq r ||h_n(t) - h(t)|| \to 0$$

when $n \to \infty$ by (H_2), from Lemma 2, we deduce that $u(t) \in D(A_{(t,h(t))}), \forall t \in I$.

Now, we are going to check that u satisfies the inclusion

$$-\frac{du}{dt}(t) \in A_{(t,h(t))} u(t) + f(t, h(t), u_h(t)) \quad \text{a.e. } t \in I$$

As $\dot{u}_{h_n} \to \dot{u}$ weakly in $L^2_H([0,1])$, $\dot{u}_{h_n} \to \dot{u}$ Komlos. Note that $f(t, h_n(t), u_{h_n}(t)) \to f(t, h(t), u(t))$ weakly in $L^2_E([0,1])$. Thus, $z_n(t) := f(t, h_n(t), u_{h_n}(t)) \to z(t) := f(t, h(t), u(t))$ Komlos. Hence, $\dot{u}_{h_n}(t) + f(t, h_n(t), u_{h_n}(t)) \to \dot{u}(t) + f(t, h(t), u(t))$ Komlos. Apply Lemma 4 to $A_{(t,h_n(t))}$ and $A_{(t,h(t))}$ to find a sequence (η_n) such that $\eta_n \in D(A_{(t,h_n(t))}), \eta_n \to \eta, A^0_{(t,h_n(t))} \eta_n \to A^0_{(t,h(t))} u(t)$. From

$$-\dot{u}_{h_n}(t) \in A_{(t,h_n(t))} u_{h_n}(t) + f(t, h_n(t), u_{h_n}(t))$$

by monotonicity

$$\langle \frac{du_{h_n}}{dt} + z_n(t), u_{h_n}(t) - \eta_n \rangle \leq \langle A^0_{(t,h_n(t))} \eta_n, \eta_n - u_{h_n}(t) \rangle.$$

From

$$\langle \frac{du_{h_n}}{dt}(t) + z_n(t), u(t) - \eta \rangle = \langle \frac{du_{h_n}}{dt}(t) + z_n(t), u_{h_n}(t) - \eta_n \rangle$$

$$+ \langle \frac{du_{h_n}}{dt}(t) + z_n(t), u(t) - u_{h_n}(t) - (\eta - \eta_n) \rangle,$$

let us write

$$\frac{1}{n} \sum_{j=1}^{n} \langle \frac{du_{h_j}}{dt}(t) + z_j(t), u(t) - \eta \rangle = \frac{1}{n} \sum_{j=1}^{n} \langle \frac{du_{h_j}}{dt}(t) + z_j(t), u_{h_j}(t) - \eta_j \rangle$$

$$+ \frac{1}{n} \sum_{j=1}^{n} \langle \frac{du_{h_j}}{dt}(t) + z_j(t), u(t) - u_{h_j}(t) \rangle$$

$$+ \sum_{j=1}^{n} \langle \frac{du_{h_j}}{dt}(t) + z_j(t), \eta_j - \eta \rangle,$$

so that

$$\frac{1}{n} \sum_{j=1}^{n} \langle \frac{du_{h_j}}{dt}(t) + z_j(t), u(t) - \eta \rangle \leq \frac{1}{n} \sum_{j=1}^{n} \langle A^0_{(t,h_j(t))} \eta_j, \eta_j - u_{h_j}(t) \rangle + (\psi(t) + M) \frac{1}{n} \sum_{j=1}^{n} ||v(t) - u_{h_j}(t)||.$$

$$+(\psi(t)+M)\frac{1}{n}\sum_{j=1}^{n}\|\eta_j-\eta\|.$$

Passing to the limit using (3) when $n \to \infty$, this last inequality gives immediately

$$\langle \frac{du}{dt}(t)+z(t), u(t)-\eta \rangle \leq \langle A^0_{(t,h(t))}\eta, \eta - u(t) \rangle \text{ a.e.}$$

As a consequence, by Lemma 1, we get $-\frac{du}{dt}(t) \in A_{(t,h(t))}u(t)+z(t)$ a.e. with $u(t) \in D(A_{(t,h(t))})$ for all $t \in [0,1]$ so that, by uniqueness, $u = u_h$.

Since $h_n \to h$, we have

$$\Phi(h_n)(t)-\Phi(h)(t) = \int_0^1 u_{h_n}(s)ds - \int_0^1 u_h(s)ds$$

$$= \int_0^1 [u_{h_n}(s) - u_h(s)]ds$$

$$\leq \int_0^1 \|u_{h_n}(s) - u_h(s)\|ds$$

As $\|u_{h_n}(\cdot) - u_h(\cdot)\| \to 0$ uniformly, we conclude that

$$\sup_{t \in [0,1]} \|\Phi(h_n)(t)-\Phi(h)(t)\| \leq \int_0^1 \|u_{h_n}(\cdot) - u_h(\cdot)\|ds \to 0$$

so that $\Phi(h_n) \to \Phi(h)$ in $\mathcal{C}_E([0,1])$. Since $\Phi : \mathcal{X}_\gamma \to \mathcal{X}_\gamma$ is continuous and $\Phi(\mathcal{X}_\gamma)$ is relatively compact in $\mathcal{C}_E(I)$, by [25,26] Φ has a fixed point, say $h = \Phi(h) \in \mathcal{X}_\gamma$ that means

$$h(t) = \Phi(h)(t) = x_0 + \int_0^t u_h(s)ds, \, t \in I,$$

$$\begin{cases} u_h(t) \in D(A_{(t,h(t))}) \\ -\frac{du_h}{dt}(t) \in A_{(t,h(t))}u_h(t) + f(t,h(t),u_h(t)) \; dt\text{-a.e.} \end{cases}$$

The proof is complete. □

4. Towards a Fractional Order of Evolution Inclusion with a Time and State Dependent Maximal Monotone Operator

Now, $I = [0,1]$ and we investigate a class of boundary value problem governed by a fractional differential inclusion (FDI) in a separable Hilbert space E coupled with an evolution inclusion governed by a time and stated dependent maximal monotone operator:

$$D^\alpha h(t) + \lambda D^{\alpha-1} h(t) = u(t), t \in I, \tag{6}$$

$$I^\beta_{0+} h(t) |_{t=0} := \lim_{t \to 0} \int_0^t \frac{(t-s)^{\beta-1}}{\Gamma(\beta)} h(s)ds = 0, \quad h(1) = I^\gamma_{0+} h(1) = \int_0^1 \frac{(1-s)^{\gamma-1}}{\Gamma(\gamma)} h(s)ds, \tag{7}$$

$$-\frac{du}{dt}(t) \in A_{(t,h(t))}u(t) \text{ a.e. } t \in I. \tag{8}$$

where $\alpha \in]1,2]$, $\beta \in [0, 2-\alpha]$, $\lambda \geq 0$, $\gamma > 0$ are given constants, D^α is the standard Riemann–Liouville fractional derivative, and Γ is the gamma function.

4.1. Fractional Calculus

For the convenience of the reader, we begin with a few reminders of the concepts that will be used in the rest of the paper.

Definition 1 (Fractional Bochner integral). *Let E be a separable Banach space. Let $f : I = [0,1] \to E$. The fractional Bochner-integral of order $\alpha > 0$ of the function f is defined by*

$$I_{a+}^{\alpha} f(t) := \int_a^t \frac{(t-s)^{\alpha-1}}{\Gamma(\alpha)} f(s) ds, \ t > a.$$

In the above definition, the sign "$\int$" denotes the classical Bochner integral.

Lemma 6 ([10]). *Let $f \in L^1([0,1], E, dt)$. We have*

(i) *If $\alpha \in]0,1[$ then $I^{\alpha} f$ exists almost everywhere on I and $I^{\alpha} f \in L^1(I, E, dt)$.*
(ii) *If $\alpha \in [1, \infty)$, then $I^{\alpha} f \in C_E(I)$.*

Definition 2. *Let E be a separable Banach space. Let $f \in L^1(I, E, dt)$. We define the Riemann–Liouville fractional derivative of order $\alpha > 0$ of f by*

$$D^{\alpha} f(t) := D_{0+}^{\alpha} f(t) = \frac{d^n}{dt^n} I_{0+}^{n-\alpha} f(t) = \frac{d^n}{dt^n} \int_0^t \frac{(t-s)^{n-\alpha-1}}{\Gamma(n-\alpha)} f(s) ds,$$

where $n = [\alpha] + 1$.

In the case $E \equiv \mathbf{R}$, we have the following well-known results.

Lemma 7 ([1,3]). *Let $\alpha > 0$. The general solution of the fractional differential equation $D^{\alpha} x(t) = 0$ is given by*

$$x(t) = c_1 t^{\alpha-1} + c_2 t^{\alpha-2} + \cdots + c_N t^{\alpha-N}, \tag{9}$$

where $c_i \in \mathbf{R}, \ i = 1, 2, \ldots, N$ (N is the smallest integer greater than or equal to α).

Remark 1. Since $D_{0+}^{\alpha} I_{0+}^{\alpha} v(t) = v(t)$, for every $v \in C(I)$, $D_{0+}^{\alpha} [I_{0+}^{\alpha} D_{0+}^{\alpha} x(t) - x(t)] = 0$ and, by Lemma 7, it follows that

$$x(t) = I_{0+}^{\alpha} D_{0+}^{\alpha} x(t) + c_1 t^{\alpha-1} + \cdots + c_N t^{\alpha-N}, \tag{10}$$

for some $c_i \in \mathbf{R}, i = 1, 2, \ldots, N$.

We denote by $W_{B,E}^{\alpha,1}(I)$ the space of all continuous functions in $C_E(I)$ such that their Riemann–Liouville fractional derivative of order $\alpha - 1$ are continuous and their Riemann–Liouville fractional derivative of order α are Bochner integrable.

4.2. Green Function and Its Properties

Let $\alpha \in]1,2], \beta \in [0, 2-\alpha], \lambda \geq 0, \gamma > 0$ and $G : [0,1] \times [0,1] \to \mathbf{R}$ be a function defined by

$$G(t,s) = \varphi(s) I_{0+}^{\alpha-1}(\exp(-\lambda t)) + \begin{cases} \exp(\lambda s) I_{s+}^{\alpha-1}(\exp(-\lambda t)), & 0 \leq s \leq t \leq 1, \\ 0, & 0 \leq t \leq s \leq 1, \end{cases} \tag{11}$$

where

$$\varphi(s) = \frac{\exp(\lambda s)}{\mu_0} \left[\left(I_{s+}^{\alpha-1+\gamma}(\exp(-\lambda t))\right)(1) - \left(I_{s+}^{\alpha-1}(\exp(-\lambda t))\right)(1) \right] \tag{12}$$

with

$$\mu_0 = \left(I_{0+}^{\alpha-1}(\exp(-\lambda t))\right)(1) - \left(I_{0+}^{\alpha-1+\gamma}(\exp(-\lambda t))\right)(1). \tag{13}$$

We recall and summarize a useful result ([28]).

Lemma 8. *Let E be a separable Banach space. Let G be the function defined by* (11)–(13).

(i) $G(\cdot,\cdot)$ satisfies the following estimate

$$|G(t,s)| \leq \frac{1}{\Gamma(\alpha)}\left(\frac{1+\Gamma(\gamma+1)}{|\mu_0|\Gamma(\alpha)\Gamma(\gamma+1)}+1\right) = M_G.$$

(ii) If $u \in W_{B,E}^{\alpha,1}([0,1])$ satisfying boundary conditions (7), then

$$u(t) = \int_0^1 G(t,s)\left(D^\alpha u(s) + \lambda D^{\alpha-1}u(s)\right)ds \quad \text{for every } t \in [0,1].$$

(iii) Let $f \in L_E^1([0,1])$ and let $u_f : [0,1] \to E$ be the function defined by

$$u_f(t) := \int_0^1 G(t,s)f(s)ds \quad \text{for} \quad t \in [0,1].$$

Then,

$$I_{0+}^\beta u_f(t)|_{t=0} = 0 \quad \text{and} \quad u_f(1) = \left(I_{0+}^\gamma u_f\right)(1).$$

Moreover $u_f \in W_{B,E}^{\alpha,1}([0,1])$ and we have

$$\left(D^{\alpha-1}u_f\right)(t) = \int_0^t \exp(-\lambda(t-s))f(s)ds + \exp(-\lambda t)\int_0^1 \varphi(s)f(s)ds \quad \text{for } t \in [0,1], \tag{14}$$

$$\left(D^\alpha u_f\right)(t) + \lambda\left(D^{\alpha-1}u_f\right)(t) = f(t) \quad \text{for all} \quad t \in [0,1]. \tag{15}$$

Remark 2. *From Lemma* 8, *we can claim that, if*

$$u_f(t) = \int_0^1 G(t,s)f(s)ds, \quad f \in L_E^1([0,1]),$$

then, for all $t \in [0,1]$,

$$\|u_f(t)\| \leq M_G\|f\|_{L_E^1([0,1])} \quad \text{and} \quad \|D^{\alpha-1}u_f(t)\| \leq M_G\|f\|_{L_E^1([0,1])}, \tag{16}$$

Indeed, by Lemma 8(i), it suffices to prove that $\|D^{\alpha-1}u_f(t)\| \leq M_G\|f\|_{L_E^1([0,1])}$. It follows from (14) that

$$\|D^{\alpha-1}u_f(t)\| \leq \int_0^1 (1+|\varphi(s)|)|f(s)|ds.$$

This, by an increase of φ (See [28] (2.9)), gives

$$\|D^{\alpha-1}u_f(t)\| \leq \Gamma(\alpha)M_G\|f\|_{L_E^1([0,1])}$$

and, since $\alpha \in [1,2]$, implies our conclusion.

4.3. Topological Structure of the Solution Set

From Lemma 8, we summarize a crucial fact.

Lemma 9. *Let E be a separable Banach space. Let $f \in L^1(I, E, dt)$. Then, the boundary value problem*

$$\begin{cases} D^\alpha u(t) + \lambda D^{\alpha-1} u(t) = f(t), & t \in I \\ I_{0+}^\beta u(t)|_{t=0} = 0, \quad u(1) = I_{0+}^\gamma u(1) \end{cases}$$

has a unique $W_{B,E}^{\alpha,1}(I)$-solution defined by

$$u(t) = \int_0^1 G(t,s) f(s) ds, \ t \in I.$$

Theorem 3. *Let E be a separable Banach space. Let $X : I \to E$ be a convex compact valued measurable multifunction such that $X(t) \subset \gamma \overline{B}_E$ for all $t \in I$, where γ is a positive constant and S_X^1 be the set of all measurable selections of X. Then, the $W_{B,E}^{\alpha,1}(I)$-solutions set of problem*

$$\begin{cases} D^\alpha u(t) + \lambda D^{\alpha-1} u(t) = f(t), f \in S_X^1, \ a.e. \ t \in I \\ I_{0+}^\beta u(t)|_{t=0} = 0, \quad u(1) = I_{0+}^\gamma u(1) \end{cases} \tag{17}$$

is compact in $C_E(I)$.

Proof. By virtue of Lemma 6, the $W_{B,E}^{\alpha,1}([0,1])$-solutions set $\mathcal{X}$ to the above inclusion is characterized by

$$\mathcal{X} = \{ u_f : I \to E, \ u_f(t) = \int_0^1 G(t,s) f(s) ds, \ f \in S_X^1, \ t \in I \}$$

Claim: $\mathcal{X}$ is bounded, convex, equicontinuous and **compact** in $C_E(I)$.

From definition of the Green function G, it is not difficult to show that $\{ u_f : f \in S_X^1 \}$ is bounded, equicontinuous in $C_E(I)$. Indeed, let $\left(u_{f_n} \right)$ be a sequence in $\mathcal{X}$. We note that, for each $n \in \mathbb{N}$, we have $u_{f_n} \in W_{B,E}^{\alpha,1}(I)$, and

$$u_{f_n}(t) = \int_0^1 G(t,s) f_n(s) ds, \quad t \in I,$$

with

- $I_{0+}^\beta u_{f_n}(t)|_{t=0} = 0, \ u_{f_n}(1) = I_{0+}^\gamma u(1),$
- $\left(D^{\alpha-1} u_{f_n} \right)(t) = \int_0^t \exp(-\lambda(t-s)) f_n(s) ds + \exp(-\lambda t) \int_0^1 \varphi(s) f_n(s) ds, \quad t \in I,$
- $\left(D^\alpha u_{f_n} \right)(t) + \lambda \left(D^{\alpha-1} u_{f_n} \right)(t) = f_n(t), t \in I.$

For $t_1, t_2 \in I$, $t_1 < t_2$, we have

$$
\begin{aligned}
u_{f_n}(t_2) - u_{f_n}(t_1) &= \int_0^1 G(t,s)(f_n(t_2,s) - f_n(t_1,s))ds \\
&= \int_0^1 \varphi(s) f_n(s) ds \left(\int_0^{t_2} \frac{e^{-\lambda \tau}}{\Gamma(\alpha-1)}(t_2-\tau)^{\alpha-2} d\tau - \int_0^{t_1} \frac{e^{-\lambda \tau}}{\Gamma(\alpha-1)}(t_1-\tau)^{\alpha-2} d\tau \right) \\
&+ \int_0^{t_2} e^{\lambda s} \left(\int_s^{t_2} \frac{(t_2-\tau)^{\alpha-2}}{\Gamma(\alpha-1)} e^{-\lambda \tau} d\tau \right) f(s) ds - \int_0^{t_1} e^{\lambda s} \left(\int_s^{t_1} \frac{e^{-\lambda \tau}}{\Gamma(\alpha-1)}(t_1-\tau)^{\alpha-2} d\tau \right) f(s) ds \\
&= \int_0^1 \phi(s) f(s) ds \left[\int_0^{t_1} e^{-\lambda \tau} \frac{(t_2-\tau)^{\alpha-2} - (t_1-\tau)^{\alpha-2}}{\Gamma(\alpha-1)} d\tau + \int_{t_1}^{t_2} e^{-\lambda \tau} \frac{(t_2-\tau)^{\alpha-2}}{\Gamma(\alpha-1)} d\tau \right] \\
&+ \int_0^{t_1} e^{\lambda s} \left(\int_s^{t_1} e^{-\lambda \tau} \frac{(t_2-\tau)^{\alpha-2} - (t_1-\tau)^{\alpha-2}}{\Gamma(\alpha-1)} d\tau \right) f(s) ds \\
&+ \int_0^{t_1} e^{\lambda s} \left(\int_{t_1}^{t_2} e^{-\lambda \tau} \frac{(t_2-\tau)^{\alpha-2}}{\Gamma(\alpha-1)} d\tau \right) f(s) ds + \int_{t_1}^{t_2} e^{\lambda s} \left(\int_s^{t_2} \frac{(t_2-\tau)^{\alpha-2}}{\Gamma(\alpha-1)} e^{-\lambda \tau} d\tau \right) f(s) ds.
\end{aligned}
$$

Then, we get

$$
\begin{aligned}
\|u_{f_n}(t_2) - u_{f_n}(t_1)\| &\leq \int_0^1 \left(|\varphi(s)| + e^{\lambda s} \right) |X(s)| ds \int_0^{t_1} e^{-\lambda \tau} \frac{(t_1-\tau)^{\alpha-2} - (t_2-\tau)^{\alpha-2}}{\Gamma(\alpha-1)} d\tau \\
&+ \int_0^1 \left(|\varphi(s)| + e^{\lambda s} \right) |X(s)| ds \int_{t_1}^{t_2} e^{-\lambda \tau} \frac{(t_2-\tau)^{\alpha-2}}{\Gamma(\alpha-1)} d\tau \\
&+ \int_{t_1}^{t_2} e^{\lambda s} |X(s)| ds \int_{t_1}^{t_2} e^{-\lambda \tau} \frac{(t_2-\tau)^{\alpha-2}}{\Gamma(\alpha-1)} d\tau.
\end{aligned}
$$

It is easy to obtain, after an integration by part, that

$$
\int_{t_1}^{t_2} e^{-\lambda \tau} \frac{(t_2-\tau)^{\alpha-2}}{\Gamma(\alpha-1)} d\tau = e^{-\lambda t_1} \frac{(t_2-t_1)^{\alpha-2}}{\Gamma(\alpha)} + \lambda \int_{t_1}^{t_2} e^{-\lambda \tau} \frac{(t_2-\tau)^{\alpha-1}}{\Gamma(\alpha)} d\tau \leq \frac{1+\lambda}{\Gamma(\alpha)}(t_2-t_1)^{\alpha-1}
$$

and

$$
\begin{aligned}
\int_0^{t_1} e^{-\lambda \tau} \frac{(t_1-\tau)^{\alpha-2} - (t_2-\tau)^{\alpha-2}}{\Gamma(\alpha-1)} d\tau &\leq \int_0^{t_1} \frac{(t_1-\tau)^{\alpha-2} - (t_2-\tau)^{\alpha-2}}{\Gamma(\alpha-1)} d\tau \\
&= \frac{(t_2-t_1)^{\alpha-1} + t_1^{\alpha-1} - t_2^{\alpha-1}}{\Gamma(\alpha)}
\end{aligned}
$$

Using the inequality that $|a^p - b^p| \leq |a-b|^p$ for all $a, b \geq 0$ and $0 < p \leq 1$, we yield

$$
\int_0^{t_1} e^{-\lambda \tau} \frac{(t_2-\tau)^{\alpha-2} - (t_1-\tau)^{\alpha-2}}{\Gamma(\alpha-1)} d\tau \leq \frac{2}{\Gamma(\alpha)}(t_2-t_1)^{\alpha-1}
$$

Then, since $\alpha \in]1, 2]$, we can increase $\|u_{f_n}(t_2) - u_{f_n}(t_1)\|$ by

$$
\|u_{f_n}(t_2) - u_{f_n}(t_1)\| \leq K|t_2 - t_1|^{\alpha-1}
$$

with $K = \int_0^1 \left[(3+\lambda)|\phi(s)| + (4+2\lambda)e^{\lambda s} \right] |X(s)| ds$ This shows that $\{u_{f_n} : n \in \mathbf{N}\}$ is equicontinuous in $C_E(I)$. Moreover, for each $t \in I$, the set $\{u_{f_n}(t) : n \in \mathbf{N}\}$ is contained in the convex compact set $\int_0^1 G(t,s) X(s) ds$ [27,29] so that $\mathcal{X}$ is relatively compact in $C_E(I)$ as claimed. Thus, we can assume that

$$
\lim_{n \to \infty} u_{f_n} = u_\infty \in C_E(I)
$$

As S_X^1 is $\sigma(L_E^1, L_{E^*}^\infty)$-compact, e.g., [29], we may assume that (f_n) $\sigma(L_E^1, L_{E^*}^\infty)$-converges to $f_\infty \in S_X^1$, so that u_{f_n} weakly converges to u_{f_∞} in $C_E(I)$ where $u_{f_\infty}(t) = \int_0^1 G(t,s)f_\infty(s)ds$ and so, for every $t \in I$,

$$u_\infty(t) = w\text{-}\lim_{n\to\infty} u_{f_n}(t) = w\text{-}\lim_{n\to\infty}\int_0^1 G(t,s)f_n(s)ds = \int_0^1 G(t,s)f_\infty(s)ds = u_{f_\infty}(t),$$

and

$$w\text{-}\lim_{n\to\infty}\left(D^{\alpha-1}u_{f_n}\right)(t) = w\text{-}\lim_{n\to\infty}\left[\int_0^t \exp(-\lambda(t-s))f_n(s)ds + \exp(-\lambda t)\int_0^1 \varphi(s)f_n(s)ds\right]$$
$$= \int_0^t \exp(-\lambda(t-s))f_\infty(s)ds + \exp(-\lambda t)\int_0^1 \varphi(s)f_\infty(s)ds$$
$$= \left(D^{\alpha-1}u_{f_\infty}\right)(t), \quad t \in I.$$

This means $u_\infty \in \mathcal{X}$, and the proof of the theorem is complete. □

Remark 3. *In the course of the proof of Theorem 3, we have proven the continuous dependence of the mappings $f \mapsto u_f$ and $f \mapsto D^{\alpha-1}u_f$ on the convex $\sigma(L_E^1, L_{E^*}^\infty)$-compact set S_X^1. This fact has some importance in further applications.*

Theorem 4. *Let $I = [0,1]$. Let $(t,x) \to A_{(t,x)} : D(A_{(t,x)}) \to 2^E$ a maximal monotone operator satisfying:*
(H_1) $\|A_{(t,x)}^0 y\| \le c(1 + \|x\| + \|y\|)$ *for all* $(t,x,y) \in I \times E \times D(A_{(t,x)})$, *for some positive constant c,*
(H_2) $dis(A_{(t,x)}, A_{(\tau,y)}) \le a(t) - a(\tau) + r\|x - y\|$, *for all* $0 \le \tau \le t \le 1$ *and for all* $(x,y) \in E \times E$, *where r is a positive number,* $a : I \to [0, +\infty[$ *is nondecreasing absolutely continuous on I with* $\dot{a} \in L^2(I, \mathbb{R}, dt)$,
(H_3) $D(A_{(t,x)}) \subset X(t) \subset \gamma\overline{B}_E$ *for all* $(t,x) \in I \times E$, *where* $X : I \to E$ *is a convex compact valued measurable mapping and γ is a positive number.*
Then, there is a $W_{B,E}^{\alpha,1}(I)$ mapping $x : I \to E$ and an absolutely continuous mapping $u : I \to E$ satisfying

$$\begin{cases} D^\alpha x(t) + \lambda D^{\alpha-1}x(t) = u(t), \ t \in I \\ I_{0^+}^\beta x(t)|_{t=0} = 0, \quad x(1) = I_{0^+}^\gamma x(1) \\ u(t) \in D(A_{(t,x(t))}) \\ -\frac{du}{dt}(t) \in A_{(t,x(t))}u(t) \text{ a.e. } t \in I. \end{cases}$$

Proof. Let us consider the convex compact subset $\mathcal{X}$ in the Banach space $\mathcal{C}_E(I)$ defined by

$$\mathcal{X} := \{u_f : I \to E : u_f(t) = \int_0^1 G(t,s)f(s)ds, \ f \in S_X^1, \ t \in I\}$$

We note that $\mathcal{X}$ is convex compact and equi-Lipschitz. Cf the proof of Theorem 3. Now, for each $h \in \mathcal{X}$, let us consider the unique absolutely continuous solution u_h to

$$\begin{cases} -\dot{u}_h(t) \in A_{(t,h(t))}u_h(t) \text{ a.e. } t \in I \\ u_h(t) \in D(A_{(t,h(t))}), \forall t \in I \\ u_h(0) = u_0 \in D(A_{(0,h(0))}) \end{cases}$$

For each h, let us set

$$\Phi(h)(t) = \int_0^1 G(t,s)u_h(s)ds, \ t \in I$$

Since $u_h(s) \in D(A_{(s,h(s))}) \subset X(s)$, then it is clear that $\Phi(h) \in \mathcal{X}$.

Now, we check that Φ is continuous. It is sufficient to show that, if (h_n) converges uniformly to h in $\mathcal{X}$, then the absolutely continuous solution u_{h_n} associated with h_n

$$\begin{cases} u_{h_n}(0) = u_0^n \in D(A_{(0,h_n(0))}) \\ u_{h_n}(t) \in D(A_{(t,h_n(t))}), \forall t \in I \\ -\dot{u}_{h_n}(t) \in A_{(t,h_n(t))} u_{h_n}(t) \quad \text{a.e. } t \in I \end{cases}$$

uniformly converges to the absolutely solution u_h associated with h

$$\begin{cases} u_h(0) = u_0 \in D(A_{(0,h(0))}) \\ u_h(t) \in D(A_{(t,h(t))}), \forall t \in [0, T] \\ -\dot{u}_h(t) \in A_{(t,h(t))} u_h(t) \quad \text{a.e. } t \in [0, T] \end{cases}$$

This fact is ensured by repeating the proof of Theorem 1. Since $h_n \to h$, we have

$$\Phi(h_n)(t) - \Phi(h)(t) = \int_0^1 G(t,s) u_{h_n}(s) ds - \int_0^1 G(t,s) u_h(s) ds$$

$$= \int_0^1 G(t,s) [u_{h_n}(s) - u_h(s)] ds$$

$$\leq \int_0^1 M_G \|u_{h_n}(s) - u_h(s)\| ds$$

As $\|u_{h_n}(\cdot) - u_h(\cdot)\| \to 0$ uniformly, we conclude that

$$\sup_{t \in I} \|\Phi(h_n)(t) - \Phi(h)(t)\| \leq \int_0^1 M_G \|u_{h_n}(\cdot) - u_h(\cdot)\| ds \to 0$$

so that $\Phi(h_n) \to \Phi(h)$ in $\mathcal{C}_E(I)$. Since $\Phi : \mathcal{X} \to \mathcal{X}$ is continuous, Φ has a fixed point, say $h = \Phi(h) \in \mathcal{X}$. This means that

$$h(t) = \Phi(h)(t) = \int_0^1 G(t,s) u_h(s) ds,$$

with

$$\begin{cases} u_h(0) \in D(A_{(0,h(0))}) \\ u_h(t) \in D(A_{(t,h(t))}), \forall t \in I \\ -\dot{u}_h(t) \in A_{(t,h(t))} u_h(t) \quad \text{a.e. } t \in I \end{cases}$$

Coming back to Lemma 9 and applying the above notations, this means that we have just shown that there exists a mapping $h \in W_E^{\alpha,\infty}(I)$ satisfying

$$\begin{cases} D^\alpha h(t) + \lambda D^{\alpha-1} h(t) = u_h(t), \\ I_{0+}^\beta h(t)|_{t=0} = 0, \quad h(1) = I_{0+}^\gamma h(1) \\ u_h(0) \in D(A_{(0,h(0))}) \\ u_h(t) \in D(A_{(t,h(t))}), \forall t \in I \\ -\dot{u}_h(t) \in A_{(t,h(t))} u_h(t) \quad \text{a.e. } t \in I \end{cases}$$

□

Now, we present an extension of the preceding theorem dealing with a Lipschitz perturbation.

Theorem 5. Let $I = [0,1]$. Let $(t,x) \to A_{(t,x)} : D(A_{(t,x)}) \to 2^E$ a maximal monotone operator satisfying:
(H_1) $||A^0_{(t,x)} y|| \leq c(1 + ||x|| + ||y||)$ for all $(t,x,y) \in I \times E \times D(A_{(t,x)})$, for some positive constant c,
(H_2) $dis(A_{(t,x)}, A_{(\tau,y)}) \leq a(t) - a(\tau) + r||x - y||$, for all $0 \leq \tau \leq t \leq 1$ and for all $(x,y) \in E \times E$, where r is a positive number, $a : I \to [0, +\infty[$ is nondecreasing absolutely continuous on I with $\dot{a} \in L^2(I, \mathbb{R}, dt)$,
(H_3) $D(A_{(t,x)}) \subset X(t) \subset \gamma \overline{B}_E$ for all $(t,x) \in I \times E$, where $X : I \to E$ is a convex compact valued measurable mapping and γ is a positive number.
Let $f : I \times E \times E \to E$ such that

(i) $f(.,x,y)$ is Lebesgue measurable on I for all $(x,y) \in E \times E$
(ii) $f(t,.,.)$ is continuous on $E \times E$,
(iii) $||f(t,x,y)|| \leq M$ for all $(t,x,y) \in I \times E \times E$,
(iv) $||f(t,x,y) - f(t,x,z)|| \leq M||y - z||$, for all $(t,x,y,z) \in I \times E \times E \times E$

for some positive constant M.
Then, there is a $W^{\alpha,1}_{B,E}(I)$ mapping $x : I \to E$ and an absolutely continuous mapping $v : I \to E$ satisfying

$$\begin{cases} D^\alpha x(t) + \lambda D^{\alpha-1} x(t) = v(t), \ t \in I \\ I^\beta_{0+} x(t)|_{t=0} = 0, \quad x(1) = I^\gamma_{0+} x(1) \\ v(t) \in D(A_{(t,x(t))}), \ t \in I \\ -\frac{dv}{dt}(t) \in A_{(t,x(t))} v(t) + f(t, x(t), v(t)) \quad a.e. \ t \in I. \end{cases}$$

Proof. Let us consider the convex compact subset $\mathcal{X}$ in the Banach space $\mathcal{C}_E(I)$ defined by

$$\mathcal{X} := \{u_f : I \to E : u_f(t) = \int_0^1 G(t,s) f(s) ds, \ f \in S^1_X, \ t \in I\}$$

We note that $\mathcal{X}$ is convex compact and equi-Lipschitz. Cf the proof of Theorem 3. Now, for each $h \in \mathcal{X}$, let us consider the unique absolutely continuous solution u_h to

$$\begin{cases} -\dot{u}_h(t) \in A_{(t,h(t))} u_h(t) + f(t, h(t), u_h(t)) \quad a.e. \ t \in I \\ u_h(t) \in D(A_{(t,h(t))}), \forall t \in I \\ u_h(0) = u_0 \in D(A_{(0,h(0))}) \end{cases}$$

Existence and uniqueness of absolutely solution u_h are ensured by the fact that the operator $B_h(t) = A_{(t,h(t))}$ is a time dependent maximal monotone operator absolutely continuous in variation (See Lemma 5), and the mapping $f_h(t,x) := f(t, h(t), y)$ is measurable with $t \in I$ and Lipschitz with $y \in E$. Furthermore, we have **the estimate** $||\dot{u}_h(t)|| \leq \psi(t)$ a.e for all $h \in \mathcal{X}$ where $\psi \in L^2(I)$ by the consideration given in Lemma 5 and the estimate of velocity given in ([22], Theorem 1). For each h, let us set

$$\Phi(h)(t) = \int_0^1 G(t,s) u_h(s) ds, \ t \in I.$$

Since $u_h(s) \in D(A_{(s,h(s))}) \subset X(s)$, then it is clear that $\Phi(h) \in \mathcal{X}$.

Now, we check that Φ is continuous. It is sufficient to show that, if (h_n) converges uniformly to h in $\mathcal{X}$, then the absolutely continuous solution u_{h_n} associated with h_n

$$\begin{cases} u_{h_n}(0) = u^n_0 \in D(A_{(0,h_n(0))}) \\ u_{h_n}(t) \in D(A_{(t,h_n(t))}), \forall t \in I \\ -\dot{u}_{h_n}(t) \in A_{(t,h_n(t))} u_{h_n}(t) + f(t, h_n(t), u_{h_n}(t)) \quad a.e. \ t \in I \end{cases}$$

uniformly converges to the absolutely solution u_h associated with h

$$\begin{cases} u_h(0) \in D(A_{(0,h(0))}) \\ u_h(t) \in D(A_{(t,h(t))}), \forall t \in I \\ -\dot{u}_h(t) \in A_{(t,h(t))} u_h(t) + f(t,h(t),u_h(t)) \text{ a.e. } t \in I \end{cases}$$

This **need careful look**. We note that u_{h_n} is equicontinuous with $||\dot{u}_{h_n}(t)|| \leq \psi(t)$ for almost all $t \in I$ and for all $n \in N$ where $\psi \in L^2$ and $u_{h_n}(t) \in D(A_{(t,h_n(t))}) \subset X(t)$ for all $t \in I$ and for all $n \in N$. Thus, by extracting subsequence, we may assume that $u_{h_n}(t) \to v(t) = v(0) + \int_0^t \dot{v}(s)ds$ with $\dot{v} \in L_E^2(I)$ for all $t \in I$ and $\dot{u}_{h_n} \to \dot{v}$ weakly in $L_E^2(I)$. Let us check that $v(t) \in D(A_{(t,h(t))})$ for all $t \in I$. We have $dis(A_{(t,h_n(t))}, A_{(t,h(t))}) \leq r||h_n(t) - h(t)|| \to 0$. It is clear that $(y_n = A_{(t,h_n(t))}^0 u_{h_n}(t))$ is bounded and hence relatively weakly compact. By applying Lemma 2 to $u_{h_n}(t) \to v(t)$ and to a convergence subsequence of (y_n) using $u_{h_n}(t) \in X(t) \subset \gamma \overline{B}_E$ to show that $v(t) \in D(A_{(t,h(t))})$. As $\dot{u}_{h_n} \to \dot{v}$ weakly in $L_E^2(I)$, $\dot{u}_{h_n} \to \dot{v}$ Komlos. Note that $f(t,h_n(t),u_{h_n}(t)) \to f(t,h(t),u_h(t))$ weakly in $L_E^2(I)$. Thus, $z_n(t) := f(t,h_n(t),u_{h_n}(t)) \to z(t) := f(t,h(t),v(t))$ Komlos. Hence, $\dot{u}_{h_n}(t) + f(t,h_n(t),u_{h_n}(t)) \to \dot{v}(t) + f(t,h(t),v(t))$ Komlos. Apply Lemma 4 to $A_{(t,h_n(t))}$ and $A_{(t,h(t))}$ to find a sequence (η_n) such that such that $\eta_n \in D(A_{(t,h_n(t))})$, $\eta_n \to \eta$, $A_{(t,h_n(t))}^0 \eta_n \to A_{(t,h(t))}^0 v(t)$ From

$$-\dot{u}_{h_n}(t) \in A_{(t,h_n(t))} u_{h_n}(t) + f(t,h_n(t),u_{h_n}(t)) (**)$$

by monotonicity

$$\langle \frac{du_{h_n}}{dt} + z_n(t), u_{h_n}(t) - \eta_n \rangle \leq \langle A_{(t,h_n(t))}^0 \eta_n, \eta_n - u_{h_n}(t) \rangle. (***)$$

From

$$\langle \frac{du_{h_n}}{dt}(t) + z_n(t), v(t) - \eta \rangle = \langle \frac{du_{h_n}}{dt}(t) + z_n(t), u_{h_n}(t) - \eta_n \rangle$$

$$+ \langle \frac{du_{h_n}}{dt}(t) + z_n(t), v(t) - u_{h_n}(t) - (\eta - \eta_n) \rangle,$$

let us write
$$\frac{1}{n} \sum_{j=1}^n \langle \frac{du_{h_j}}{dt}(t) + z_j(t), v(t) - \eta \rangle =$$

$$\frac{1}{n} \sum_{j=1}^n \langle \frac{du_{h_j}}{dt}(t) + z_j(t), u_{h_j}(t) - \eta_j \rangle + \frac{1}{n} \sum_{j=1}^n \langle \frac{du_{h_j}}{dt}(t) + z_j(t), v(t) - u_{h_j}(t) \rangle$$

$$+ \sum_{j=1}^n \langle \frac{du_{h_j}}{dt}(t) + z_j(t), \eta_j - \eta \rangle,$$

so that

$$\frac{1}{n} \sum_{j=1}^n \langle \frac{du_{h_j}}{dt}(t) + z_j(t), v(t) - \eta \rangle \leq \frac{1}{n} \sum_{j=1}^n \langle A_{(t,h_j(t))}^0 \eta_j, \eta_j - u_{h_j}(t) \rangle + (\psi(t) + M) \frac{1}{n} \sum_{j=1}^n ||v(t) - u_{h_j}(t)||.$$

$$+ (\psi(t) + M) \frac{1}{n} \sum_{j=1}^n ||\eta_j - \eta||.$$

Passing to the limit using (5) when $n \to \infty$, this last inequality gives immediately

$$\langle \frac{dv}{dt}(t) + z(t), v(t) - \eta \rangle \leq \langle A_{(t,h(t))}^0 \eta, \eta - v(t) \rangle \text{ a.e.}$$

As a consequence, by Lemma 1, we get
$-\frac{dv}{dt}(t) \in A_{(t,h(t))}v(t) + z(t)$ a.e. with $v(t) \in D(A_{(t,h(t))})$ for all $t \in I$ so that, by uniqueness, $v = u_h$. Since $h_n \to h$, we have

$$\Phi(h_n)(t) - \Phi(h)(t) = \int_0^1 G(t,s)u_{h_n}(s)ds - \int_0^1 G(t,s)u_h(s)ds$$

$$= \int_0^1 G(t,s)[u_{h_n}(s) - u_h(s)]ds$$

$$\leq \int_0^1 M_G||u_{h_n}(s) - u_h(s)||ds$$

As $||u_{h_n}(\cdot) - u_h(\cdot)|| \to 0$ uniformly, we conclude that

$$\sup_{t \in I} ||\Phi(h_n)(t) - \Phi(h)(t)|| \leq \int_0^1 M_G||u_{h_n}(\cdot) - u_h(\cdot)||ds \to 0$$

so that $\Phi(h_n) \to \Phi(h)$ in $\mathcal{C}_E(I)$. Since $\Phi : \mathcal{X} \to \mathcal{X}$ is continuous, Φ has a fixed point, say $h = \Phi(h) \in \mathcal{X}$. This means that

$$h(t) = \Phi(h)(t) = \int_0^1 G(t,s)u_h(s)ds,$$

with

$$\begin{cases} u_h(0) \in D(A_{(0,h(0))}) \\ u_h(t) \in D(A_{(t,h(t))}), \forall t \in I \\ -\dot{u}_h(t) \in A_{(t,h(t))}u_h(t) + f(t,h(t),u_h(t)) \quad \text{a.e. } t \in I \end{cases}$$

Coming back to Lemma 9 and applying the above notations, this means that we have just shown that there exists a mapping $h \in W_{B,E}^{\alpha,\infty}(I)$ satisfying

$$\begin{cases} D^{\alpha}h(t) + \lambda D^{\alpha-1}h(t) = u_h(t), \\ I_{0^+}^{\beta}h(t)|_{t=0} = 0, \quad h(1) = I_{0^+}^{\gamma}h(1) \\ u_h(0) \in D(A_{(0,h(0))}) \\ u_h(t) \in D(A_{(t,h(t))}), \forall t \in I \\ -\dot{u}_h(t) \in A_{(t,h(t))}u_h(t) + f(t,h(t),u_h(t)) \quad \text{a.e. } t \in I \end{cases}$$

□

We finish the paper by investigating a fractional order to a sweeping process [30,31].

We begin recall the existence of absolutely continuous solution to a class of sweeping process [18,32].

Theorem 6. *Let $f : [0,T] \to E$ be a continuous mapping such that $||f(t)|| \leq \beta$ for all $t \in [0,T]$, let $v : [0,T] \to \mathbf{R}^+$ be a positive nondecreasing continuous function with $v(0) = 0$. Let $C : [0,T] \to E$ be a convex weakly compact valued mapping such that $d_H(C(t), C(\tau)) \leq |v(t) - v(\tau)|$ for all $t, \tau \in [0,T]$. Let $A : E \to E$ be a linear continuous coercive symmetric operator and let $B : E \to E$ be a linear continuous compact operator. Then, for any $u_0 \in E$, the evolution inclusion*

$$f(t) + Bu(t) - A\frac{du}{dt}(t) \in N_{C(t)}(\frac{du}{dt}(t))$$

$$u(0) = u_0$$

admits a unique $W_E^{1,\infty}([0,T])$ solution $u : [0,T] \to E$.

Theorem 7. Let $f : I \times E \to E$ be a bounded continuous mapping such that $\|f(t,x)\| \leq M$ for all $(t,x) \in I \times E$, for some positive constant M, let $v : I \to \mathbf{R}^+$ be a positive nondecreasing continuous function with $v(0) = 0$. Let $C : I \to E$ be a convex compact valued mapping such that $d_H(C(t), C(\tau)) \leq |v(t) - v(\tau)|$ for all $t, \tau \in I$. Let $A : E \to E$ be a linear continuous coercive symmetric operator and let $B : E \to E$ be a linear continuous compact operator.

Then, for any $u_0 \in E$, there exists a $W^{\alpha,1}_{B,E}(I)$ mapping $x : I \to E$ and an absolutely continuous mapping $u : I \to E$ satisfying

$$\begin{cases} u(0) = u_0 \in E \\ D^\alpha x(t) + \lambda D^{\alpha-1} x(t) = u(t), \ t \in I \\ I^\beta_{0+} x(t)|_{t=0} = 0, \ x(1) = I^\gamma_{0+} x(1) \\ f(t,x(t)) + Bu(t) - A\frac{du}{dt}(t) \in N_{C(t)}(\frac{du}{dt}(t)), \text{ a.e. } t \in I \end{cases}$$

Proof. By Theorem 6 and the assumptions on f, for any bounded continuous mapping $h : I \to E$, there is a unique absolutely continuous solution v_h to the inclusion

$$\begin{cases} v_h(0) = u_0 \in E \\ f(t,h(t)) + Bv_h(t) - A\frac{dv_h}{dt}(t) \in N_{C(t)}(\frac{dv_h}{dt}(t)), \text{ a.e. } t \in I \end{cases}$$

with $\frac{dv_h}{dt}(t) \in C(t)$ a.e. so that $v_h(t) = u_0 + \int_0^t \frac{dv_h}{ds}(s)ds \in u_0 + \int_0^t C(s)ds, \forall t \in I$. By our assumption, C is scalarly upper semicontinuous convex compact valued integrably bounded: $C(t) \subset \rho \overline{B}_E, \forall t \in I$, hence, by [33], $t \mapsto \Psi(t) := u_0 + \int_0^t C(s)ds$ is a scalarly upper semicontinuous convex **compact** valued integrably bounded mapping with $\Psi(t) := u_0 + \int_0^t C(s)ds \subset u_0 + \rho \overline{B}_E, \forall t \in I$. Let us consider the closed convex subset $\mathcal{X}$ in the Banach space $\mathcal{C}_E(I)$ defined by

$$\mathcal{X} := \{u_f : I \to E : u_f(t) = \int_0^1 G(t,s)f(s)ds, \ f \in S^1_{u_0 + \rho \overline{B}_E}, \ t \in I\},$$

where $S^1_{u_0 + \rho \overline{B}_E}$ denotes the set of all integrable selections of the convex weakly compact valued constant multifunction $u_0 + \rho \overline{B}_E$. Now, for each $h \in \mathcal{X}$, let us consider the mapping defined by

$$\Phi(h)(t) := \int_0^t G(t,s)v_h(s)ds,$$

for $t \in I$. Then, it is clear that $\Phi(h) \in \mathcal{X}$. Since $u_0 + \int_0^t C(s)ds$ is a convex compact, $\Phi(\mathcal{X})$ is equicontinuous and relatively compact in the Banach space $\mathcal{C}_E(I)$ by virtue of Theorem 3 using the compactness of $\Psi(t)$. Now, we check that Φ is continuous. It is sufficient to show that, if (h_n) uniformly converges to h in $\mathcal{X}$, then the absolutely continuous solution v_{h_n} associated with h_n

$$\begin{cases} v_{h_n}(0) = u_0 \in E \\ f(t,h_n(t)) + Bv_{h_n}(t) - A\frac{dv_{h_n}}{dt}(t) \in N_{C(t)}(\frac{dv_{h_n}}{dt}(t)), \text{ a.e. } t \in I \end{cases}$$

uniformly converges to the absolutely continuous solution v_h associated with h

$$\begin{cases} v_h(0) = u_0 \in E \\ f(t,h(t)) + Bv_h(t) - A\frac{dv_h}{dt}(t) \in N_{C(t)}(\frac{dv_h}{dt}(t)), \text{ a.e. } t \in I \end{cases}$$

As (v_{h_n}) is equi-absolutely continuous with $v_{h_n}t) \in u_0 + \int_0^t C(s)ds, \forall t \in I$, we may assume that (v_{h_n}) uniformly converges to an absolutely continuous mapping z.

Since $v_{h_n}(t) = u_0 + \int_{]0,t]} \frac{dv_{h_n}}{ds}(s)ds$, $t \in I$ and $\frac{dv_{h_n}}{ds}(s) \in C(s)$, a.e. $s \in I$, we may assume that $(\frac{dv_{h_n}}{dt})$ weakly converges in $L_E^1(I)$ to $w \in L_E^1(I)$ with $w(t) \in C(t)$, $t \in I$ so that

$$\lim_n v_{h_n}(t) = u_0 + \int_0^t w(s)ds := u(t), \ t \in I.$$

By identifying the limits, we get

$$u(t) = z(t) = u_0 + \int_0^t w(s)ds$$

with $\dot{u} = w$. Therefore, by applying the arguments in the variational limit result in [34], we get

$$f(t,h(t)) + Bu(t) - A\frac{du}{dt}(t)) \in N_{C(t)}(\frac{du}{dt}(t)), \text{ a.e. } t \in I$$

with $u(0) = u_0 \in E$, so that, by uniqueness, $u = v_h$. Since $h_n \to h$, we have

$$\Phi(h_n)(t) - \Phi(h)(t) = \int_0^1 G(t,s)v_{h_n}(s)ds - \int_0^1 G(t,s)v_h(s)ds$$

$$= \int_0^1 G(t,s)[v_{h_n}(s) - v_h(s)]ds$$

$$\leq \int_0^1 M_G \|v_{h_n}(s) - v_h(s)\| ds$$

As $\|v_{h_n}(\cdot) - v_h(\cdot)\| \to 0$ uniformly, we conclude that

$$\sup_{t \in I} \|\Phi(h_n)(t) - \Phi(h)(t)\| \leq \int_0^1 M_G \|v_{h_n}(\cdot) - v_h(\cdot)\| ds \to 0$$

so that $\Phi(h_n) \to \Phi(h)$ in $\mathcal{C}_E(I)$. Since $\Phi : \mathcal{X} \to \mathcal{X}$ is continuous and $\Phi(\mathcal{X})$ is relatively compact in $\mathcal{C}_E(I)$, by [25,26] Φ has a fixed point, say $h = \Phi(h) \in \mathcal{X}$. This means that

$$h(t) = \Phi(h)(t) = \int_0^1 G(t,s)v_h(s)ds,$$

with

$$\begin{cases} v_h(0) = u_0 \in E \\ D^\alpha h(t) + \lambda D^{\alpha-1}h(t) = v_h(t), \ t \in I \\ I_{0^+}^\beta h(t)|_{t=0} = 0, \ h(1) = I_{0^+}^\gamma h(1) \\ f(t,h(t)) + Bv_h(t) - A\frac{dv_h}{dt}(t)) \in N_{C(t)}(\frac{dv_h}{dt}(t)), \text{ a.e. } t \in I \end{cases}$$

The proof is complete. □

Theorem 8. *Theorems 6 and 7 results are inspired by some ideas in [18]. At this point, some variants are available, mainly when the second member is a time dependent subdifferential operator [35], namely, for any $u_0 \in E$, there exists a $W_{B,E}^{\alpha,1}(I)$ mapping $x : I \to E$ and an absolutely continuous mapping $u : I \to E$ satisfying*

$$\begin{cases} u(0) = u_0 \in E \\ D^\alpha x(t) + \lambda D^{\alpha-1}x(t) = u(t), \ t \in I \\ I_{0^+}^\beta x(t)|_{t=0} = 0, \ x(1) = I_{0^+}^\gamma x(1) \\ f(t,x(t)) + Bu(t) - A\frac{du}{dt}(t) \in \partial\varphi(t,\frac{du}{dt}(t)), \text{ a.e. } t \in I \end{cases}$$

5. On a Fillipov Theorem

We end this section with a Fillipov theorem and a relaxation theorem for the fractional differential inclusion

$$\begin{cases} D^\alpha u(t) + \lambda D^{\alpha-1} u(t) \in F(t, u(t)), a.e.\ t \in I \\ I_{0^+}^\beta u(t)|_{t=0} = 0, \quad u(1) = I_{0^+}^\gamma u(1) \end{cases}$$

where $F : I \times E \to E$ is a closed valued Lipschitz mapping w.r.t.o $x \in E$.

Theorem 9. *Assume that E is a separable Banach space. Let $F : I \times E \to E$ be a closed valued $\mathcal{L}(I) \otimes \mathcal{B}(E)$-measurable mapping such that*
($\mathcal{H}_1$): $d_H(F(t,x), F(t,y)) \leq l(t)\|x - y\|$ for all t, x, y where $l \in L^1_\mathbf{R}(I)$) such that $\rho := M_G \|l\|_{L^1_\mathbf{R}(I)} < 1$.
Assume further that
($\mathcal{H}_2$) : there exists $g \in L^1_E(I)$ such that $d(g(t), F(t, u_g(t))) < \frac{l(t)}{\sum_{n=1}^\infty n\rho^{n-1}}$ where $u_g(t) = \int_0^1 G(t,s) g(s) ds, \forall t \in I$.
Then, the fractional differential inclusion

$$\begin{cases} D^\alpha u(t) + \lambda D^{\alpha-1} u(t) \in F(t, u(t)), a.e.\ t \in I \\ I_{0^+}^\beta u(t)|_{t=0} = 0, \quad u(1) = I_{0^+}^\gamma u(1) \end{cases}$$

has at least a $W_{B,E}^{\alpha,1}(I)$-solution $u : I \to E$.

Proof. We use the ideas in the proof of Theorem 4.3 in [36], Remark 2 and Lemma 9.
It is worth mentioning that the series $\Lambda := \sum_{n=1}^\infty n\rho^{n-1}$ is convergent. Indeed, we have

$$\lim_{n \to \infty} \frac{(n+1)\rho^n}{n\rho^{n-1}} = \lim_{n \to \infty} \frac{n+1}{n} \rho = \rho < 1.$$

Thus, by d'Alembert's ratio test, the series $\sum_{n=1}^\infty n\rho^{n-1}$ is convergent

Step 1. We shall construct inductively sequence $\{f_n(\cdot)\}_{n=1}^\infty$ where $f_1 = g$ such that the following conditions are fulfilled, for all $n \geq 1$,

$$f_n \in L^1_E(I) \quad \text{and} \quad f_{n+1}(t) \in F(t, u_{f_n}(t)), t \in I, \tag{18}$$

$$\|f_{n+1}(t) - f_n(t)\| \leq (n+1)\rho^{n-1} l(t) \Lambda^{-1}, \tag{19}$$

$$\left\| u_{f_{n+1}}(t) - u_{f_n}(t) \right\| = \left\| \int_0^1 G(t,s)[f_{n+1}(s) - f_n(s)] ds \right\| \leq (n+1)\rho^n \Lambda^{-1}, \tag{20}$$

for all $t \in I$. We note that the passage from (18) to (19) is obtained, thanks to (16) of Remark 2, with

$$\left\| u_{f_{n+1}}(t) - u_{f_n}(t) \right\| = \left\| \int_0^1 G(t,s)[f_{n+1}(s) - f_n(s)] ds \right\| < M_G \|f_{n+1}(t) - f_n(t)\|$$

By $(\mathcal{H}_2)$, we have $d(f_1(t), F(t, u_{f_1}(t))) < l(t)\Lambda^{-1}$, $t \in I$. Let us consider the multifunction $\Sigma_1 : I \to c(E)$ defined by

$$\Sigma_1(t) = \left\{ v \in F(t, u_{f_1}(t)) : \|v - f_1(t)\| \leq 2l(t)\Lambda^{-1} \right\}.$$

Clearly, Σ_1 is Lebesgue measurable with nonempty closed values. In view of the existence theorem of measurable selections (see [29]), there is a measurable function $f_2 : I \to E$ such that $f_2(t) \in \Sigma_1(t)$ for all $t \in I$. This yields

$$f_2(t) \in F(t, u_{f_1}(t)), \quad \|f_2(t) - f_1(t)\| \leq 2l(t)\Lambda^{-1},$$

for all $t \in I$. Thus, it is easy to see that $f_2 \in L^1_E(I)$ and

$$\left\| u_{f_2}(t) - u_{f_1}(t) \right\| = \left\| \int_0^1 G(t,s)[f_2(s) - f_1(s)]ds \right\| \leq 2\rho\Lambda^{-1},$$

for all $t \in I$.

- Suppose that we have constructed integrable functions $f_1, f_2, \ldots, f_n$ such that

$$f_{i+1}(t) \in F(t, u_{f_i}(t)), \ t \in I,$$

$$\|f_{i+1}(t) - f_i(t)\| \leq (i+1)\rho^{i-1}l(t)\Lambda^{-1},$$

for all $i = 1, 2, \ldots, n-1$. Then,

$$\left\| u_{f_{i+1}}(t) - u_{f_i}(t) \right\| = \left\| \int_0^1 G(t,s)[f_{i+1}(s) - f_i(s)]ds \right\| \leq (i+1)\rho^i\Lambda^{-1},$$

for $i = 1, 2, \ldots, n-1$.

- The function f_{n+1} is constructed as follows. We have

$$d\left(f_n(t), F\left(t, u_{f_n}(t)\right)\right) \leq d_H\left(F(t, u_{f_{n-1}}(t)), F(t, u_{f_n}(t))\right)$$
$$\leq l(t)\left\| u_{f_n}(t) - u_{f_{n-1}}(t) \right\|$$
$$\leq n\rho^{n-1}l(t)\Lambda^{-1}.$$

The multifunction $\Sigma_n : I \to c(E)$, defined by

$$\Sigma_n(t) = \left\{ v \in F(t, u_n(t)) : \|v - f_n(t)\| \leq (n+1)\rho^{n-1}l(t)\varepsilon\Lambda^{-1} \right\},$$

is Lebesgue measurable with nonempty closed values. Thus, there exists a measurable function f_{n+1} such that

$$f_{n+1}(t) \in F\left(t, u_{f_n}(t)\right), \quad \|f_{n+1}(t) - f_n(t)\| \leq (n+1)\rho^{n-1}l(t)\Lambda^{-1},$$

for all $t \in I$. Then, it is clear that, for all $t \in I$,

$$\left\| u_{f_{n+1}}(t) - u_{f_n}(t) \right\| = \left\| \int_0^1 G(t,s)[f_{n+1}(s) - f_n(s)]ds \right\| \leq (n+1)\rho^n\Lambda^{-1},$$

Thus, such a sequence $\{f_n\}_{n=1}^\infty$ with the required properties exists.

Step 2. It follows that, for all $n \geq 1$, we have

$$\|f_{n+1} - f_n\|_{L^1_E(I)} = \int_0^1 \|f_{n+1}(t) - f_n(t)\| \, dt \leq (n+1)\rho^{n-1} \|l\|_{L^1_{\mathbb{R}^+}(I)} \Lambda^{-1}. \tag{21}$$

On the other hand, by $\rho < 1$ the series $\sum_{n=1}^{\infty}(n+1)\rho^{n-1}$ is convergent (using d'Alembert's ratio test). Now, we assert that $\{f_n(\cdot)\}_{n=1}^{\infty}$ is a Cauchy sequence in $L_E^1(I)$. Indeed, using (10), for $n, m \in \mathbf{N}$ such that $m > n$, we have the estimate

$$\|f_m - f_n\|_{L_E^1(I)} \le \|f_{n+1} - f_n\|_{L_E^1(I)} + \|f_{n+2} - f_{n+1}\|_{L_E^1(I)} + \cdots + \|f_m - f_{m-1}\|_{L_E^1(I)}$$

$$\le \left[(n+1)\rho^{n-1} + (n+2)\rho^n + \cdots + m\rho^{m-2}\right] \|l\|_{L_{\mathbf{R}_+}^1(I)} \Lambda^{-1}$$

$$\le \left(\sum_{k=n}^{\infty}(k+1)\rho^{k-1}\right) \|l\|_{L_{\mathbf{R}_+}^1(I)} \Lambda^{-1}$$

Letting $n \to \infty$ in the above inequality, we see that $\|f_m - f_n\|_{L_E^1(I)}$ goes to 0 when m, n goes to ∞. Since the normed space $L_E^1(I)$ is complete, (f_n) norm converges to an element $f \in L_E^1(I)$. By the properties of our Green function and the definition of u_{f_n}, we conclude that u_{f_n} pointwise converge with respect to the norm topology to u_f

$$u_f(t) = \int_0^1 G(t,s) f(s) ds, \forall t \in I.$$

Now, we claim that $f(t) \in F(t, u_f(t))$, a.e. $t \in I$. Let us write

$$d(f(t), F(t, u_f(t))) \le \left| d(f(t), F(t, u_f(t))) - d(f_n(t), F(t, u_f(t))) \right|$$
$$+ d(f_n(t), F(t, u_f(t))). \tag{22}$$

On the other hand,

$$\left| d(f(t), F(t, u_f(t))) - d(f_n(t), F(t, u_f(t))) \right| \le \|f(t) - f_n(t)\|, \tag{23}$$

and, by $f_n(t) \in F(t, u_{f_{n-1}}(t))$, $t \in I$, we have

$$d(f_n(t), F(t, u_f(t))) \le d_H\left(F(t, u_{f_{n-1}}(t)), F(t, u_f(t))\right)$$
$$\le l(t) \left\| u_{f_{n-1}}(t) - u_f(t) \right\|. \tag{24}$$

Since $(f_n)_{n \in \mathbf{N}}$ norm converges to $f \in L_E^1(I)$, we may, by extracting subsequences, assume that $\|f_n(t) - f(t)\|_E \to 0$ a.e. Now, passing to the limit when $n \to \infty$ in the preceding inequality, we get

$$d(f(t), F(t, u_f(t))) = 0 \quad a.e. \ t \in I$$

This implies that $f(t) \in F(t, u_f(t))$, a.e.$t \in I$ because F is closed valued. Thus, by Lemma 9, we have shown that u_f is a solution of the problem

$$\begin{cases} D^\alpha u_f(t) + \lambda D^{\alpha-1} u_f(t) \in F(t, u_f(t)), a.e. \ t \in I \\ I_{0+}^\beta u_f(t)|_{t=0} = 0, \quad u_f(1) = I_{0+}^\gamma u_f(1) \end{cases}$$

The proof of theorem is complete.
□

A relaxation theorem is available using the machinery developed in [36] Theorem 4.2 and Lemma 9.

Theorem 10. Relaxation Assume that E is a separable Banach space. Let $F : I \times E \to E$ be a closed valued $\mathcal{L}(I) \otimes \mathcal{B}(E)$-measurable mapping such that

($\mathcal{H}_1$): $d_H(F(t,x), F(t,y)) \leq l(t)||x-y||$ for all t, x, y where $l \in L^1_{\mathbb{R}}(I)$) such that $\rho := M_G ||l||_{L^1_{\mathbb{R}}(I)} < 1$.
Assume further that
($\mathcal{H}_2$) : there exists $g \in L^1_E(I)$ such that $d(g(t), F(t, u_g(t))) < \frac{l(t)}{\sum_{n=1}^{\infty} n\rho^{n-1}}$ where $u_g(t) = \int_0^1 G(t,s)g(s)ds, \forall t \in I$.
($\mathcal{H}_3$) : $d(0, F(t,x)) < c(t)(1+||x||), \forall (t,x) \in I \times E$ where c is a positive integrable function.
Then, the following holds:
(a)

$$(\mathcal{P}_F) \begin{cases} D^\alpha u(t) + \lambda D^{\alpha-1} u(t) \in F(t, u(t)), a.e.\ t \in I \\ I^\beta_{0+} u(t)|_{t=0} = 0, \quad u(1) = I^\gamma_{0+} u(1) \end{cases}$$

and

$$(\mathcal{P}_{\overline{co}F}) \begin{cases} D^\alpha u(t) + \lambda D^{\alpha-1} u(t) \in \overline{co}F(t, u(t)), a.e.\ t \in I \\ I^\beta_{0+} u(t)|_{t=0} = 0, \quad u(1) = I^\gamma_{0+} u(1) \end{cases}$$

have at least a solution in $W^{\alpha,1}_{B,E}(I)$.
(b) Let $f_0 \in L^1_E(I))$ such that

$$f_0(t) \in \overline{co}F(t, u_{f_0}(t))$$

$$u_{f_0}(t) = \int_0^1 G(t,s) f_0(s)ds, \forall t \in I$$

Then, for every $\varepsilon > 0$, there exists $f \in L^1_E(I)$ such that

$$f(t) \in F(t, u_f(t)), \quad a.e.$$

$$u_f(t) = \int_0^1 G(t,s) f(s)ds, \forall t \in I$$

and

$$\sup_{t \in I} ||u_f(t) - u_{f_0}(t)|| \leq \varepsilon.$$

Proof. We will proceed in several steps.

Step 1. (a) follows from Theorem 9 applied to both F and $\overline{co}F$ taking account of ($\mathcal{H}_1$) − ($\mathcal{H}_2$). Let $u_{f_0}(\cdot)$ be a $W^{\alpha,1}_{B,E}(I)$-solution of the problem ($\mathcal{P}_{\overline{co}F}$) that is, $u_{f_0} \in \mathcal{S}_{\mathcal{P}_{\overline{co}F}}$

$$f_0(t) \in \overline{co}F(t, u_{f_0}(t)), \ a.e.\ t \in I, \tag{25}$$

$$u_{f_0}(t) := \int_0^1 G(t,s) f_0(s)ds, \forall t \in I \tag{26}$$

Let S^1_F and $S^1_{\overline{co}F}$ denote the set of all $L^1_E(I)$-selections of the set valued mappings $t \to F(t, u_{f_0}(t))$ and $t \mapsto \overline{co}F(t, u_{f_0}(t))$ By ($\mathcal{H}_3$), the multifunction $t \mapsto F(t, u_{f_0}(t))$ is closed valued and integrable:

$$d(0, F(t, u_{f_0}(t))) < c(t)(1+||u_{f_0}(t)||)$$

so that S^1_F is non empty. Then, according to Hiai–Umegaki [37], $S^1_{\overline{co}F} = \overline{co}S^1_F$ where $\overline{co}$ is taken in $L^1_E(I)$. This equality along with $f_0(t) \in \overline{co}F(t, u_{f_0}(t))$, a.e. $t \in I$ yields $f_0 \in \overline{co}S^1_F$. Let $\varepsilon > 0$. There exists $g_\varepsilon \in L^1_E(I)$ such that $g_\varepsilon \in coS^1_F$ and $||f_0 - g_\varepsilon||_{L^1_E(I)} \leq \frac{1}{2}\varepsilon\Lambda^{-1}M_G^{-1}$ so that

$$||u_{f_0}(t) - u_{g_\varepsilon}(t)|| < \frac{1}{2}\varepsilon\Lambda^{-1}.$$

As $g_\varepsilon \in coS_F^1$, then $g_\varepsilon = \sum_{i=1}^n \lambda_i f_I$ with $f_i \in L_E^1(I)$, $f_i(t) \in F(t, u_{f_0}(t))$, $\lambda_i \geq 0$, $\sum_{i=1}^n \lambda_i = 1$. Let $\Phi(t) := \{f_i(t : 1 \leq i \leq n\}$, then $\Phi(t)$ is a compact valued integrably bounded mapping with $|\Phi(t)| \leq r(t) := \sup_{1 \leq i \leq n} |f_i(t)|$. Then, from [38], there exists

$$h_1 \in L_E^1(I), h_1(t) \in \Phi(t) \subset F(t, u_{f_0}(t))), \forall t \in I$$

such that

$$\sup_{0 \leq t < \tau \leq 1} \left\| \int_t^\tau [h_1(s) - g_\varepsilon(s)] ds \right\| \leq \frac{1}{2} \varepsilon M_G^{-1} \Delta^{-1}$$

so that

$$\|u_{h_1}(t) - u_{g_\varepsilon}(t)\| = \left\| \int_0^1 G(t,s)[h_1(s) - g_\varepsilon(s)] ds \right\| \leq M_G \left\| \int_0^1 [h_1(s) - g_\varepsilon(s)] ds \right\| \leq \frac{1}{2} \varepsilon \Delta^{-1}.$$

Consequently,

(*) $$\|u_{h_1}(t) - u_{f_0}(t)\| \leq \varepsilon \Delta^{-1}.$$

Step 2. We shall construct inductively sequence $\{h_n(\cdot)\}_{n=1}^\infty$ such that the following conditions are fulfilled, for all $n \geq 1$,

$$h_n \in L_E^1(I) \quad \text{and} \quad h_{n+1}(t) \in F(t, u_{h_n}(t)), t \in I, \tag{27}$$

$$\|h_{n+1}(t) - h_n(t)\| \leq (n+1)\rho^{n-1} l(t) \varepsilon \Lambda^{-1}, \tag{28}$$

$$\left\| u_{h_{n+1}}(t) - u_{h_n}(t) \right\| = \left\| \int_0^1 G(t,s)[h_{n+1}(s) - h_n(s)] ds \right\| \leq (n+1)\rho^n \varepsilon \Lambda^{-1}, t \in I \tag{29}$$

- The multifunction $F(\cdot, u_{h_1}(\cdot))$ is Lebesgue-measurable and

$$d_H\left(F(t, u_{h_1}(t)), F(t, u_{f_0}(t))\right) \leq l(t) \left\| u_{h_1}(t) - u_{f_0}(t) \right\|$$

This implies that, for $t \in I$,

$$d_H\left(F(t, u_{h_1}(t)), F(t, u_{f_0}(t))\right) \leq l(t) \varepsilon \Lambda^{-1},$$

As $h_1(t) \in F(t, u_{f_0}(t))$, we have $d(h_1(t), F(t, u_{h_1}(t))) \leq l(t) \varepsilon \Lambda^{-1}$, $t \in I$. Let us consider the multifunction $\Sigma_1 : I \to c(E)$ defined by

$$\Sigma_1(t) = \left\{ v \in F(t, u_{h_1}(t)) : \|v - h_1(t)\| \leq 2l(t) \varepsilon \Lambda^{-1} \right\}.$$

Clearly, Σ_1 is Lebesgue measurable with nonempty closed values. In view of the existence theorem of measurable selections (see [29]), there is a measurable function $h_2 : I \to \bar{E}$ such that $h_2(t) \in \Sigma_1(t)$ for all $t \in I$. This yields

$$h_2(t) \in F(t, u_{h_1}(t)), \quad \|h_2(t) - h_1(t)\| \leq 2l(t) \varepsilon \Lambda^{-1},$$

for all $t \in I$. Thus, it is easy to see that $h_2 \in L_E^1(I)$ and

$$\|u_{h_2}(t) - u_{h_1}(t)\| = \left\| \int_0^1 G(t,s)[h_2(s) - h_1(s)] ds \right\| \leq 2\rho\varepsilon \Lambda^{-1},$$

for all $t \in I$.

- Suppose that we have constructed integrable functions $h_1, h_2, \ldots, h_n$ such that

$$h_{i+1}(t) \in F(t, u_{h_i}(t)), \text{ a.e.t} \in I,$$

$$\|h_{i+1}(t) - h_i(t)\| \leq (i+1)\rho^{i-1}l(t)\varepsilon\Lambda^{-1},$$

for all $i = 1, 2, \ldots, n-1$. Then,

$$\left\|u_{h_{i+1}}(t) - u_{h_i}(t)\right\| = \left\|\int_0^1 G(t,s)[h_{i+1}(s) - h_i(s)]ds\right\| \leq (i+1)\rho^i\varepsilon\Lambda^{-1},$$

for $i = 1, 2, \ldots, n-1$.

♦ The function h_{n+1} is constructed as follows. We have

$$d(h_n(t), F(t, u_{h_n}(t))) \leq d_H(F(t, u_{h_{n-1}}(t)), F(t, u_{h_n}(t)))$$

$$\leq l(t)\|u_{h_n}(t) - u_{h_{n-1}}(t)\| \leq n\rho^{n-1}l(t)\varepsilon\Lambda^{-1}$$

The multifunction $\Sigma_n : I \to c(E)$, defined by

$$\Sigma_n(t) = \left\{v \in F(t, u_{h_n}(t)) : \|v - h_n(t)\| \leq (n+1)\rho^{n-1}l(t)\varepsilon\Lambda^{-1}\right\},$$

is Lebesgue measurable with nonempty closed values. Thus, there exists a measurable function h_{n+1} such that

$$h_{n+1}(t) \in F(t, u_{h_n}(t)), \quad \|h_{n+1}(t) - h_n(t)\| \leq (n+1)\rho^{n-1}l(t)\varepsilon\Lambda^{-1},$$

for all $t \in I$. Then, it is clear that, for all $t \in I$,

$$\left\|u_{h_{n+1}}(t) - u_{h_n}(t)\right\| = \left\|\int_0^1 G(t,s)[h_{n+1}(s) - h_n(s)]ds\right\| \leq (n+1)\rho^n\varepsilon\Lambda^{-1},$$

Thus, a sequence $\{h_n\}_{n=1}^\infty$ satisfying (27)–(29) exists.

Step 3. It follows from (28) that, for all $n \geq 1$, we have

$$\|h_{n+1} - h_n\|_{L^1_E(I)} = \int_0^1 \|h_{n+1}(t) - h_n(t)\|\,dt \leq (n+1)\rho^{n-1}\|l\|_{L^1_{\mathbf{R}_+}(I)}\varepsilon\Lambda^{-1}. \quad (30)$$

On the other hand, by $\rho < 1$, the series $\sum_{n=1}^\infty (n+1)\rho^{n-1}$ is convergent (using d'Alembert's ratio test). Now, we assert that $\{h_n(\cdot)\}_{n=1}^\infty$ is a Cauchy sequence in $L^1_E(I)$. Indeed, using (30), for $n, m \in \mathbf{N}$, such that $m > n$, we have the estimate

$$\|h_m - h_n\|_{L^1_E(I)} \leq \|h_{n+1} - h_n\|_{L^1_E(I)} + \|h_{n+2} - h_{n+1}\|_{L^1_E(I)} + \cdots + \|h_m - h_{m-1}\|_{L^1_E(I)}$$

$$\leq \left[(n+1)\rho^{n-1} + (n+2)\rho^n + \cdots + m\rho^{m-2}\right]\|l\|_{L^1_{\mathbf{R}_+}(I)}\varepsilon\Lambda^{-1}$$

$$\leq \left(\sum_{k=n}^\infty (k+1)\rho^{k-1}\right)\|l\|_{L^1_{\mathbf{R}_+}(I)}\varepsilon\Lambda^{-1}$$

Letting $n \to \infty$ in the above inequality, we see that $\|h_m - h_n\|_{L^1_E(I)}$ goes to 0 when m, n goes to ∞. Since the normed space $L^1_E(I)$ is complete, (h_n) norm converges to an element $f \in L^1_E(I)$. By the properties of our Green function and the definition of u_{h_n}, we conclude that u_{h_n} pointwise converges with respect to the norm topology to u_f where

$$u_f(t) = \int_0^1 G(t,s)f(s)ds.$$

Moreover, from (29), we deduce that

$$\left\|u_{h_n}(t) - u_{f_0}(t)\right\| \leq \|u_{h_1}(t) - u_{f_0}(t)\| + \|u_{h_2}(t) - u_{h_1}(t)\| + \ldots + \|u_{h_n}(t) - u_{h_{n-1}}(t)\|$$

$$\leq \left(\sum_{j=1}^{n} j\rho^{j-1}\right)\varepsilon\Lambda^{-1}$$

for all $t \in I$. Recall that $\Lambda = \sum_{n=1}^{\infty} n\rho^{n-1}$. Thus, by letting $n \to \infty$ in the last inequality, we get

$$\left\|u_f - u_{f_0}\right\|_{C_E(I)} = \max_{t \in I}\left\|u_f(t) - u_{f_0}(t)\right\| \leq \varepsilon.$$

Now, we claim that $f(t) \in F(t, u_f(t))$, a.e. $t \in I$. Let us write

$$d(f(t), F(t, u_f(t))) \leq \left|d(f(t), F(t, u_f(t))) - d(h_n(t), F(t, u_f(t)))\right|$$

$$+ d(h_n(t), F(t, u_f(t))). \tag{31}$$

On the other hand,

$$\left|d(f(t), F(t, u_f(t))) - d(h_n(t), F(t, u_f(t)))\right| \leq \|f(t) - h_n(t)\|, \tag{32}$$

and, by $h_n(t) \in F(t, u_{h_{n-1}}(t))$, $t \in I$, we have

$$d(h_n(t), F(t, u_f(t))) \leq d_H\left(F(t, u_{h_{n-1}}(t)), F(t, u_f(t))\right)$$

$$\leq l(t) \left\|u_{h_{n-1}}(t) - u_f(t)\right\| \tag{33}$$

Since $(h_n)_{n \in \mathbb{N}}$ norm converges to $f \in L^1_E(I)$ we may, by extracting subsequences, assume that $\|h_n(t) - f(t)\|_E \to 0$ a.e. Now, passing to the limit when $n \to \infty$ in (31)–(33), we get

$$d(f(t), F(t, u_f(t))) = 0 \quad a.e. \, t \in I$$

This implies that $f(t) \in F(t, u_f(t))$, a.e. $t \in I$ because F is closed valued. Hence, u_f is a solution of the problem $(\mathcal{P}_F)$, satisfying the required density property. The proof of theorem is complete. □

6. Conclusions

In the context of separable Hilbert space, our algorithm and tools are fairly general and they allow for treating several variants of system of fractional differential inclusion coupled with a time and state dependent maximal monotone operators with Lipschitz perturbation, in particular the second order solution of evolution inclusion governed time and state dependent maximal monotone operators with Lipschitz perturbation. Our results contain novelties. Nevertheless, there are several issues—for instance, the existence of solutions for the case of closed unbounded Lipschitz perturbation that is needed in the optimal control.

Author Contributions: All authors have contributed equally to this work for writing, review and editing. All authors have read and agreed to the published version of the manuscript.

Funding: This research received no external funding.

Conflicts of Interest: The authors declare no conflict of interest.

References

1. Kilbas, A.A.; Srivastava, H.M.; Trujillo, J.J. *Theory and Applications of Fractional Differential Equations*; Math. Studies 204; Elsevier: Amsterdam, The Netherlands, 2006.
2. Miller, K.S.; Ross, B. *An Introduction to the Fractional Calculus and Fractional Differential Equations*; Wiley: New York, NY, USA, 1993.
3. Podlubny, I. *Fractional Differential Equations*; Academic Press: New York, NY, USA, 1999.
4. Samko, S.G.; Kilbas, A.A.; Marichev, O.I. *Fractional Integrals and Derivatives: Theory and Applications*; Gordon and Breach: New York, NY, USA, 1993.
5. Vrabie, I.L. *Compactness Methods for Nonlinear Evolution Equations, Pitman Monographs and Surveys in Pure and Applied Mathematics (Vol. 32), Longman Scientific and Technical*; John Wiley: New York, NY, USA, 1987.
6. Graef, J.R.; Henderson, J.; Ouahab, A. *Impulsive Differential Inclusions. A Fixed Point Approach*; De Gruyter Series in Nonlinear Analysis and Applications 20; de Gruyter: Berlin, Germany, 2013.
7. Zhou, Y. *Basic Theory of Fractional Differential Equations*; World Scientific Publishing: Singapore, 2014.
8. Ahmad, B.; Nieto, J. Riemann–Liouville fractional differential equations with fractional boundary conditions. *Fixed Point Theory* **2012**, *13*, 329–336.
9. Ouahab, A. Some results for fractional boundary value problem of differential inclusions. *Nonlinear Anal.* **2008**, *69*, 3877–3896.
10. Phung, P.H.; Truong, L.X. On a fractional differential inclusion with integral boundary conditions in Banach space. *Fract. Calc. Appl. Anal.* **2013**, *16*, 538–558.
11. El-Sayed, A.M.A.; Ibrahim, A.G. Set-valued integral equations of arbitrary (fractional) order. *Appl. Math. Comput.* **2001**, *118*, 113–121.
12. Castaing, C.; Truong, L.X.; Phung, P.D. On a fractional differential inclusion with integral boundary condition in Banach spaces. *J. Nonlinear Convex Anal.* **2016**, *17*, 441–471.
13. Benchohra, M.; Graef, J.; Mostefai, F.Z. Weak solutions for boundary-value problems with nonlinear fractional differential inclusions. *Nonlinear Dyn. Syst. Theory* **2011**, *11*, 227–237.
14. El-Sayed, A.M.A. Nonlinear functional differential equations of arbitrary orders. *Nonlinear Anal.* **1998**, *33*, 181–186.
15. Cernea, A. On a fractional differential inclusion with boundary condition. *Stud. Univ. Babes-Bolyai Math.* **2010**, *LV*, 105–113.
16. Agarwal, R.P.; Arshad, S.; O'Reagan, D.; Lupulescu, V. Fuzzy Fractional Integral Equations under Compactness type condition. *Frac. Calc. Appl. Anal.* **2012**, *15*, 572–590.
17. Benchohra, M.; Henderson, J.; Ntouyas, S.K.; Ouahab, A. Existence results for fractional functional differential inclusions with infinite delay and applications to control theory. *Fract. Calc. Appl. Anal.* **2008**, *11*, 35–56.
18. Castaing, C.; Monteiro Marques, M.D.P.; Saidi, S. Evolution problems with time-dependent subdifferential operators. *Adv. Math. Econ.* **2020**, *23*, 1–39.
19. Barbu, V. *Nonlinear Semigroups and Differential Equations in Banach Spaces*; Noordhoff International Publishing: Leyden, The Netherlands, 1976.
20. Brezis, H. *Opérateurs Maximaux Monotones*; Elsevier: Amsterdam, The Netherlands, 1973.
21. Vladimirov, A.A. Nonstationnary dissipative evolution equation in Hilbert space. *Nonlinear Anal.* **1991**, *17*, 499–518.
22. Azzam-Laouir, D.; Belhoula, W.; Castaing, C.; Monteiro Marques, M.D.P. Multi-valued perturbation to evolution problems involving time dependent maximal monotone operators. *Evol. Equ. Control Theory* **2020**, *9*, 219–254.
23. Kunze, M.; Monteiro Marques, M.D.P. BV solutions to evolution problems with time-dependent domains. *Set-Valued Anal.* **1997**, *5*, 57–72.
24. Castaing, C.; Ibrahim A.G.; Yarou, M. Some contributions to nonconvex sweeping process. *J. Nonlinear Convex Anal.* **2009**, *10*, 1–20.
25. Idzik, A. Almost fixed points theorems. *Proc. Am. Math. Soc.* **1988**, *104*, 779–784.
26. Park, S. Fixed points of approximable or Kakutani maps. *J. Nonlinear Convex Anal.* **2006**, *7*, 1–17.
27. Castaing, C. Quelques résultats de compacité liés a l'intégration. *CR Acad. Sci. Paris* **1970**, *270*, 1732–1735; reprinted in *Bull. Soc. Math. France* **1972**, *31*, 73–81.

28. Castaing, C.; Godet-Thobie, C.; Phung, P.D.; Truong, L.X. On fractional differential inclusions with nonlocal boundary conditions. *Fract. Calc. Appl. Anal.* **2019**, *22*, 444–478.
29. Castaing, C.; Valadier, M. *Convex Analysis and Measurable Multifunctions*; Lecture Notes in Mathematics, 580; Springer: Berlin/Heidelberg, Germany, 1977.
30. Monteiro Marques, M.D.P. Differential Inclusions Nonsmooth Mechanical Problems, Shocks and Dry Friction. In *Progress in Nonlinear Differential Equations and Their Applications*; Springer: Berlin, Germany, 1993; Volume 9.
31. Moreau, J.J. Evolution problem asssociated with a moving convex set in a Hilbert space. *J. Differ. Equ.* **1977**, *26*, 347–374.
32. Adly, S.; Haddad, T.; Thibault, L. Convex sweeping process in the framework of measure differential inclusions and evolution variational inequalities. *Math. Program* **2014**, *148*, 5–47.
33. Castaing, C. Weak compactness and convergences in Bochner and Pettis integration. *Vietnam J. Math.* **1996**, *24*, 241–286.
34. Castaing, C.; Marques, M.D.P.; Raynaud de Fitte, P. Second order evolution problems with time dependent maximal monotone operator. *Adv. Math. Econ.* **2018**, *23*, 25–77
35. Kenmochi, N. Solvability of nonlinear evolution equations with time-dependent constraints and applications. *Bull. Fac. Educ. Chiba Univ.* **1981**, *30*, 1.
36. Castaing, C.; Truong, L.X. Some Topological Properties of Solution Sets in a Second Order Differential Inclusion with m-point Boundary Conditions. *Set-Valued Var. Anal.* **2012**, *20*, 249–277.
37. Hiai, F.; Umegaki, H. Integrals, conditional expectations, and martingales of multivalued functions. *J. Multivar. Anal.* **1977**, *7*, 149–182.
38. Tolstonogov, A.A.; Tolstonogov, D. Lp Continuous Extreme Selectors of Multifunctions with Decomposable Values: Relaxation Theorems. *Set-Valued Anal.* **1996**, *4*, 237–269.

© 2020 by the authors. Licensee MDPI, Basel, Switzerland. This article is an open access article distributed under the terms and conditions of the Creative Commons Attribution (CC BY) license (http://creativecommons.org/licenses/by/4.0/).

Article

Inequalities in Triangular Norm-Based ∗-Fuzzy $(L^+)^p$ Spaces

Abbas Ghaffari [1], Reza Saadati [2,*] and Radko Mesiar [3,4,*]

[1] Department of Mathematics, Science and Research Branch, Islamic Azad University, Tehran 1477893855, Iran; a.g139571@gmail.com
[2] School of Mathematics, Iran University of Science and Technology, Narmak, Tehran 1311416846, Iran
[3] Department of Algebra and Geometry, Faculty of Science, Palacký University Olomouc, 17. listopadu 12, 771 46 Olomouc, Czech Republic
[4] Department of Mathematics Radlinského 11, Faculty of Civil Engineering, 810 05 Bratislava, Slovakia
* Correspondence: rsaadati@eml.cc or rsaadati@iust.ac.ir (R.S.); mesiar@math.sk (R.M.)

Received: 19 October 2020; Accepted: 3 November 2020; Published: 6 November 2020

Abstract: In this article, we introduce the ∗-fuzzy $(L^+)^p$ spaces for $1 \leq p < \infty$ on triangular norm-based ∗-fuzzy measure spaces and show that they are complete ∗-fuzzy normed space and investigate some properties in these space. Next, we prove Chebyshev's inequality and Hölder's inequality in ∗-fuzzy $(L^+)^p$ spaces.

Keywords: fuzzy measure space; fuzzy integration; t-norm; Chebyshev's inequality; Hölder's inequality

MSC: Primary 54C40, 14E20; Secondary 46E25, 20C20

Function spaces, especially L^p spaces, play an important role in many parts in analysis. The impact of L^p spaces follows from the fact that they offer a partial but useful generalization of the fundamental L^1 space of integrable functions. The standard analysis, based on sigma-additive measures and Lebesgue–Stieltjess integral, including also several integral inequalities, has been generalized in the past decades into set-valued analysis, including set-valued measures, integrals, and related inequalities. Some subsequent generalizations are based on fuzzy sets [1,2] and include fuzzy measures, fuzzy integrals and several fuzzy integral inequalities. Our aim is the further development of fuzzy set analysis, expanding our original proposal given in [3]. In fact, we use a new model of the fuzzy measure theory (∗-fuzzy measure) which is a dynamic generalization of the classical measure theory. Our model of the fuzzy measure theory created by replacing the non-negative real range and the additivity of classical measures with fuzzy sets and triangular norms. Moreover, the ∗-fuzzy measure theory has been motivated by defining new additivity property using triangular norms. Our approach is related to the idea of fuzzy metric spaces [4–7] and can be apply for decision making problems [8,9].

In this paper, we shall work on a fixed triangular norm-based ∗-fuzzy measure space $(X, \mathcal{C}, \mu, *)$ introduced in [3] which was derived from the idea of fuzzy and probabilistic metric spaces [5–7,10,11]. Using the concept of fuzzy measurable functions and fuzzy integrable functions we define a special class of function spaces named by ∗-fuzzy $(L^+)^p$. After some overview given in Sections 2–4 and devoted to the basic information concerning ∗-fuzzy measures and related integration, in Section 5 we define a norm on ∗-fuzzy $(L^+)^p$ spaces and show these spaces are complete ∗-fuzzy normed space in the sense of Cheng-Mordeson and others [12–15]. This definition of ∗-fuzzy norm helps us to prove Chebyshev's Inequality and Hölder's Inequality.

1. ∗–Fuzzy Measure

First, we recall some basic concepts and notations that will be used throughout the paper. Let X be a non-empty set, $\mathcal{C}$ be a σ-algebra of subsets of X. Unless stated otherwise, all subsets of X are supposed to belong to $\mathcal{C}$. Here, we let $I = [0,1]$.

Definition 1. *([10,11]) A continuous triangular norm (shortly, a ct-norm) is a continuous binary operation $*$ from $I^2 = [0,1]^2$ to I such that*

(a) $\varsigma * \tau = \tau * \varsigma$ and $\varsigma * (\tau * v) = (\varsigma * \tau) * v$ for all $\varsigma, \tau, v \in [0,1]$;
(b) $\varsigma * 1 = \varsigma$ for all $\varsigma \in I$;
(c) $\varsigma * \tau \leq v * \iota$ whenever $\varsigma \leq v$ and $\tau \leq \iota$ for all $\varsigma, \tau, v, \iota \in I$.

Some examples of the *ct*-norms are as follows.

1. $\varsigma *_P \tau = \varsigma \tau$ (: the product *t*-norm);
2. $\varsigma *_M \tau = \min\{\varsigma, \tau\}$ (: the minimum *t*-norm);
3. $\varsigma *_L \tau = \max\{\varsigma + \tau - 1, 0\}$ (: the Lukasiewicz *t*-norm);
4.

$$\varsigma *_H \tau = \begin{cases} 0, & \text{if } \varsigma = \tau = 0, \\ \dfrac{1}{\frac{1}{\varsigma} + \frac{1}{\tau} - 1}, & \text{otherwise,} \end{cases}$$

(: the Hamacher product *t*-norm).

We define

$$*_{i=1}^{k} \varsigma_i = \varsigma_1 * \varsigma_2 * \cdots * \varsigma_k,$$

for $k \in \{2, 3, \cdots\}$, which is well defined due to the associativity of the operation $*$. Moreover,

$$*_{i=1}^{\infty} \varsigma_i = \lim_{k \to \infty} *_{i=1}^{k} \varsigma_i,$$

which is well defined due to the monotonicity and boundedness of the operation $*$.

Now, we introduce the concept of $*$-fuzzy measure.

Definition 2 ([3]). *Let X be a set and $\mathcal{C}$ be a σ-algebra consisting of subsets of X. A fuzzy measure on $\mathcal{C} \times (0, \infty)$ is a fuzzy set $\mu : \mathcal{C} \times (0, \infty) \to I$ such that*

(i) $\mu(\varnothing, \tau) = 1, \ \forall \tau \in (0, \infty)$;
(ii) *if* $A_i \in \mathcal{C}, i = 1, 2, \cdots$, *are pairwise disjoint, then*

$$\mu(\cup_{i=1}^{\infty} A_i, \tau) = *_{i=1}^{\infty} \mu(A_i, \tau), \ \forall \tau \in (0, \infty).$$

Saying the A_i are pairwise disjoint means that $A_i \cap A_j = \varnothing$, if $i \neq j$.

Definition 2 is known as countable $*$-additivity. We say a fuzzy measure μ is finitely $*$-additive if, for any $n \in \mathbb{N}$

$$\mu(\cup_{i=1}^{n} A_i, \tau) = *_{i=1}^{n} \mu(A_i, \tau), \ \forall \tau \in (0, \infty).$$

whenever $A_1, \cdots, A_n$ are in $\mathcal{C}$ and are pairwise disjoint. The quadruple $(X, \mathcal{C}, \mu, *)$ is called a $*$-fuzzy measure space (in short, $*$-FMS).

Example 1. *Let $(X, \mathcal{C}, m)$ be a measurable space. Let $* = *_H$ and define*

$$\mu_0(A, \tau) = \frac{\tau}{\tau + m(A)}, \quad \forall \tau \in (0, \infty),$$

then $(X, \mathcal{C}, \mu_0, *)$ is a $*$-FMS.

Example 2. *Let $(X, \mathcal{C}, m)$ be a measurable space. Let $* = *_P$. Define*

$$\mu_0(A, \tau) = e^{-\frac{m(A)}{\tau}}, \quad \forall \tau \in (0, \infty).$$

Then, μ_0 is a $*$-FM on $\mathcal{C} \times (0, \infty)$.

2. $*$-Fuzzy Measurable Functions

Now, we review the concept of $*$-fuzzy normed spaces, for more details, we refer to the works in [12–15].

Definition 3. *Let X be a vector space, $*$ be a ct-norm and the fuzzy set N on $X \times (0, \infty)$ satisfies the following conditions for all $x, y \in X$ and $\tau, \sigma \in (0, \infty)$,*

(i) $N(x, \tau) > 0$.
(ii) $N(x, \tau) = 1 \Leftrightarrow x = 0$.
(iii) $N(\alpha x, \tau) = N\left(x, \frac{\tau}{|\alpha|}\right)$ for every $\alpha \neq 0$.
(iv) $N(x, \tau) * N(y, \sigma) \leq N(x + y, \tau + \sigma)$.
(v) $N(x, .) : (0, \infty) \to (0, 1]$ is continuous.
(vi) $\lim_{\tau \to \infty} N(x, \tau) = 1$ and $\lim_{\tau \to 0} N(x, \tau) = 0$.

Then, N is called a $*$-fuzzy norm on X and $(X, N, *)$ is called $*$-fuzzy normed space.

Assume that $(\mathbb{R}, |.|)$ is a standard normed space, we define: $N(x, \tau) = \frac{\tau}{\tau + |x|}$ with $* = *_P$, it is obvious $(\mathbb{R}, N, *_P)$ is a $*$-fuzzy normed space.

Let $(X, N, *)$ be a $*$-fuzzy normed space. We define the open ball $\mathcal{B}(x, r, \tau)$ and the closed ball $\mathcal{B}[x, r, \tau]$ with center $x \in X$ and radius $0 < r < 1$, $\tau > 0$ as follows,

$$\mathcal{B}(x, r, \tau) = \{y \in X : N(x - y, \tau) > 1 - r\}, \quad (1)$$
$$\mathcal{B}[x, r, \tau] = \{y \in X : N(x - y, \tau) \geq 1 - r\}. \quad (2)$$

Let $(X, N, *)$ be a $*$-fuzzy normed space. A set $E \subset X$ is said to be open if for each $x \in E$, there is $0 < r_x < 1$ and $\tau_x > 0$ such that $\mathcal{B}(x, r_x, \tau_x) \subseteq E$. A set $F \subseteq X$ is said to be closed in X in case its complement $F^c = X - F$ is open in X.

Let $(X, N, *)$ be a $*$-fuzzy normed space. A subset $E \subseteq X$ is said to be fuzzy bounded if there exist $\tau > 0$ and $r \in (0, 1)$ such that $N(x - y, \tau) > 1 - r$ for all $x, y \in E$.

Let $(X, N, *)$ be a $*$-fuzzy normed space. A sequence $\{x_n\} \subset X$ is fuzzy convergent to an $x \in X$ in $*$-fuzzy normed space $(X, N, *)$ if for any $\tau > 0$ and $\epsilon > 0$ there exists a positive integer $N_\epsilon > 0$ such that $N(x_n - x, \tau) > 1 - \epsilon$ whenever $n \geq N_\epsilon$.

Now, we define $*$-fuzzy measurable functions.

Definition 4. *Let $(X, \mathcal{C})$ and $(Y, \mathcal{D})$ be $*$-fuzzy measurable spaces. A mapping $f : X \to Y$ is called $*$-fuzzy $(\mathcal{C}, \mathcal{D})$-measurable if $f^{-1}(E) \in \mathcal{C}$ for all $E \in \mathcal{D}$. If X is any $*$-fuzzy normed space, the σ-algebra generated by* the family of open sets in X (or, equivalently, by the family of closed sets in X) is called the Borel σ-algebra on X and is denoted by $\mathcal{B}_X$.

3. *-Fuzzy Integration

In this section, we recall the concept of *-fuzzy integration by using fuzzy simple functions on the *-FMS $(X, \mathcal{C}, *, \mu)$ and add some new results.

Definition 5. *Let $(X, \mathcal{C}, *, \mu)$ be *-FMS, we define*

$$L_+ = \{f : X \to [0, \infty) \mid f \text{ is fuzzy } (\mathcal{C}, \mathcal{B}_\mathbb{R})\text{-measurable function}\}.$$

If ϕ is a simple fuzzy $((\mathcal{C}, \mathcal{B}_\mathbb{R})$-measurable) function in L_+ with standard representation $\phi = \sum_{i=1}^n a_i \chi_{E_i}$, where $a_i > 0$ and $E_i \in \mathcal{C}$ for $i = 1, ..., n$, and $E_i \cap E_j = \emptyset$ for $i \neq j$, we define the fuzzy integral of ϕ as

$$\int_X \phi(x) d\mu(x, \tau) = \int_X \sum_{i=1}^n a_i \chi_{E_i} d\mu(x, \tau) = *_{i=1}^n \mu\left(E_i, \frac{\tau}{a_i}\right).$$

In [3], the authors have shown that, with respect to $\mu(A, \tau)$, μ satisfies the following statement;

(i) $\mu : (A, .) : (., \infty) \to [0, 1]$ is increasing and continuous.
(ii) $\mu\left(A, \frac{\tau}{a+b}\right) \geq \mu\left(A, \frac{\tau}{a}\right) * \mu\left(A, \frac{\tau}{b}\right)$ for every $a, b > 0$, $\tau \in (0, \infty)$.
(iii) $\lim_{\tau_n \to \tau_0}\left(*_{i=1}^k \mu(A_i, \tau_n)\right) = *_{i=1}^k \lim_{\tau_n \to \tau_0} \mu(A_i, \tau_n)$ for every $A_i \cap A_j = \emptyset$.
(iv) $\lim_{\tau \to 0} \mu(E, \tau) = 0$ and $\lim_{\tau \to \infty} \mu(E, \tau) = 1$.
(v) $\lim_{\tau_n \to \tau_0} \lim_{m \to \infty}\left(\mu\left(E^m, \frac{\tau_n}{a^m}\right)\right) = \lim_{m \to \infty} \lim_{\tau_n \to \tau_0}\left(\mu\left(E^m, \frac{\tau_n}{a^m}\right)\right)$.

If $A \in \mathcal{C}$, then $\phi \chi_A$ is also fuzzy simple function $\left(\phi \chi_A = \sum_{i=1}^n a_i \chi_{A \cap E_i}\right)$, and we define $\int \phi(x) d\mu(x, \tau)$ to be $\int \phi \chi_A d\mu(x, \tau)$.

Theorem 1 ([3])**.** *Let ϕ and ψ be simple functions in L_+. Then, we have*

(i) $\int_X 0 d\mu(x, \tau) = 1$.
(ii) *If $c \in (0, 1]$ then $\int_X (c\phi)(x) d\mu(x, \tau) \geq c \int_X \phi(x) d\mu(x, \tau)$, and for $c \in [1, \infty)$ we have $\int_X (c\phi)(x) d\mu(x, \tau) \leq c \int_X \phi(x) d\mu(x, \tau)$, $\forall \tau \in (0, \infty)$.*
(iii) *If $\phi \leq \psi$, then $\int_X \phi(x) d\mu(x, \tau) \geq \int_X \psi(x) d\mu(x, \tau)$.*
(iv) *The map $A \to \int_A \phi(x) d\mu(x, \tau)$ is a fuzzy measure on $\mathcal{C}$, $\forall \tau \in (0, \infty)$.*

In the next theorem, we prove an important fuzzy integral inequality for fuzzy simple functions.

Theorem 2. *Let ϕ and ψ be fuzzy simple functions in L_+, then*

$$\int (\phi + \psi)(x) d\mu(x, \tau) \geq \left(\int \phi(x) d\mu(x, \tau)\right) * \left(\int \psi(x) d\mu(x, \tau)\right).$$

Proof. Let ϕ and ψ be fuzzy simple functions in L_+, then we have

$$\int_X (\phi + \psi)(x) d\mu(x, \tau), \tag{3}$$

$$= \int_X \left(\left(\sum_{i=1}^n a_i \chi_{E_i}(x) \right) + \left(\sum_{j=1}^m b_j \chi_{F_j}(x) \right) \right) d\mu(x, \tau),$$

$$= \int_X \left(\sum_{i,j} (a_i + b_j) \chi_{E_i \cap F_j}(x) \right) d\mu(x, \tau),$$

$$= *_{i=1}^n *_{j=1}^m \mu \left((E_i \cap F_j), \frac{\tau}{(a_i + b_j)} \right).$$

On the other hand,

$$\left(\int_X \phi(x) d\mu(x, \tau) * \int_X \psi(x) d\mu(x, \tau) \right), \tag{4}$$

$$= \left(\int_X \left(\sum_{i=1}^n a_i \chi_{E_i}(x) \right) d\mu(x, \tau) \right) * \left(\int_X \left(\sum_{j=1}^m b_j \chi_{F_j}(x) \right) d\mu(x, \tau) \right),$$

$$= \left(*_{i=1}^n *_{j=1}^m \mu \left((E_i \cap F_j), \frac{\tau}{a_i} \right) \right) * \left(*_{j=1}^m *_{i=1}^n \mu \left((E_i \cap F_j), \frac{\tau}{b_j} \right) \right),$$

$$= *_{i=1}^n *_{j=1}^m \left(\mu \left((E_i \cap F_j), \frac{\tau}{a_i} \right) * \mu \left((E_i \cap F_j), \frac{\tau}{b_j} \right) \right),$$

$$\leq *_{i=1}^n *_{j=1}^m \left(\mu \left((E_i \cap F_j), \frac{\tau}{(a_i + b_j)} \right) \right).$$

From (3) and (4), we get

$$\int_X (\phi + \psi)(x) d\mu(x, \tau) \geq \left(\int_X \phi(x) d\mu(x, \tau) \right) * \left(\int_X \psi(x) d\mu(x, \tau) \right).$$

□

Now, we extend the concept of fuzzy integral to all functions in L_+.

Definition 6. *Let f be a fuzzy measurable function in L_+, we define fuzzy integral by*

$$\int_X f(x) d\mu(x, \tau)$$

$$= \inf \left\{ \int_X \phi(x) d\mu(x, \tau) \mid 0 \leq \phi \leq f, \ \phi \text{ is fuzzy simple function} \right\}.$$

By Theorem 1 (iii), the two definitions of $\int f$ agree when f is fuzzy simple function, as the family of fuzzy simple functions over which the infimum is taken includes f itself. Moreover, it is obvious from the definition that $\int f \geq \int g$ whenever $f \leq g$, and $\int cf \geq c \int f$ for all $c \in (0, 1]$ and $\int cf \leq c \int f$ for all $c \in [1, \infty)$ and $\int (f + g) \geq (\int f) * (\int g)$.

Definition 7. *If $f \in L_+$, we say that f is fuzzy integrable if $\int f d\mu(x, \tau) > 0$ for each $\tau > 0$. Let $(X, \mathcal{C}, \mu, *)$ be a $*$-FMS. We define*

$$L^+ := \left\{ f : X \to [0, \infty), f \text{ is measurable function and } \int f(x) d\mu(x, \tau) > 0 \right\}.$$

Theorem 3 ([3]). **(The fundamental convergence theorem).** Let $(X, \mathcal{C}, \mu, *)$ be a $*$-FMS. Let f_n be a sequence in L^+ such that $f_n \longrightarrow f$ almost everywhere, then $f \in L^+$ and $\int f = \lim\limits_{n \longrightarrow \infty} \int f_n$.

$*$-Fuzzy L^+ Spaces

Here, we are ready to show that every L^+ is a $*$-fuzzy normed space. It is clear if we define

$$L := \{f : X \longrightarrow \mathbb{R}, \ f \text{ is fuzzy measurable function}\},$$

then $(L, +, .)_\mathbb{R}$ is a vector space. Moreover, in [3] the authors proved that if $f, g \in L^+$, then $|f - g| \in L^+$. Using definition L and L^+ we can show $L^+ \subseteq L$. In L^+ we define $f \leq g$ if and only if $f(x) \leq g(x)$ and so $(L^+, \leq)$ is a cone.

Note. Recall that, due to the continuity of t-norm $*$, for any systems $\{a_n\}_{n \in \mathbb{N}}$ and $\{b_n\}_{n \in \mathbb{N}}$ of elements form I we have $\inf\{a_n * b_n\} = \inf\{a_n\} * \inf\{b_n\}$.

In the next theorem we define a fuzzy norm on L^+ and prove that $(L^+, N, *)$ is a $*$-fuzzy normed space.

Theorem 4. Let $N : L^+ \times (0, \infty) \longrightarrow (0, 1]$ be a fuzzy set, such that $N(f, \tau) = \int f d\mu(x, \tau)$, then $(L^+, N, *)$ is a $*$-fuzzy normed space.

Proof.

(FN1) $N(f, \tau) = \int f d\mu(x, \tau) > 0$.

(FN2) By theorem 4.5 of [3] we have

$$N(f, \tau) = 1 \iff \int f d\mu(x, \tau) = 1 \iff f = 0$$

almost everywhere.

(FN3) Let $f = \phi = \sum\limits_{i=1}^{n} a_i \chi_{E_i}$ and $c > 0$ so,

$$N(c\phi, \tau) = \int c\phi d\mu(x, \tau), \tag{5}$$

$$= \int \sum_{i=1}^{n} a_i \chi_{E_i} d\mu(x, \tau),$$

$$= *_{i=1}^{n} \mu\left(E_i, \frac{\tau}{ca_i}\right).$$

On the other hand,

$$N\left(\phi, \frac{\tau}{c}\right) = \int \phi d\mu\left(x, \frac{\tau}{c}\right), \tag{6}$$

$$= \int \sum_{i=1}^{n} a_i \chi_{E_i} d\mu\left(x, \frac{\tau}{c}\right),$$

$$= *_{i=1}^{n} \mu\left(E_i, \frac{\tau}{ca_i}\right).$$

From (5) and (6) we conclude that

$$N(c\phi, \tau) = N\left(\phi, \frac{\tau}{c}\right). \tag{7}$$

Now, if $f \in L^+$ we have $\{\phi_n\} \subseteq L^+$ such that $\phi_n \uparrow f$, then $c\phi_n \uparrow cf$ so

$$\int c\phi_n d\mu(x,\tau) \downarrow \int cf d\mu(x,\tau).$$

By (7), we have $\int c\phi_n d\mu(x,\tau) = \int \phi_n d\mu(x, \frac{\tau}{c})$, and so

$$\int \phi_n d\mu(x, \frac{\tau}{c}) \downarrow \int cf d\mu(x,\tau). \tag{8}$$

On the other hand,

$$\int \phi_n d\mu(x, \frac{\tau}{c}) \downarrow \int f d\mu(x, \frac{\tau}{c}), \tag{9}$$

by (8) and (9) we have,

$$\int cf d\mu(x,\tau) = \int f d\mu(x, \frac{\tau}{c}),$$

$$N(cf, \tau) = N(f, \frac{\tau}{c}).$$

(FN4) Let $f = \sum_{i=1}^{m} a_i \chi_{E_i}$, $g = \sum_{j=1}^{n} b_j \chi_{F_j}$ then,

$$N(\phi + \psi, s + \tau) = \int (\phi + \psi) d\mu(x, \tau + s),$$
$$= \int \sum_{i,j} (a_i + b_j) \chi_{E_i \cap F_j} d\mu(x, \tau + s),$$
$$= *_{i,j} \mu \left(E_i \cap F_j, \frac{\tau + s}{a_i + b_j} \right).$$

On the other hand

$$N(\phi, s) * N(\psi, \tau) = \left(\int \phi d\mu(x,s) \right) * \left(\int \psi d\mu(x,\tau) \right), \tag{10}$$
$$= \left(\int \sum_{i,j} a_i \chi_{E_i \cap F_j} d\mu(x,s) \right) * \left(\int \sum_{i,j} b_j \chi_{E_i \cap F_j} d\mu(x,\tau) \right),$$
$$= \left(*_{i,j} \mu(E_i \cap F_j, \frac{s}{a_i}) \right) * \left(*_{i,j} \mu(E_i \cap F_j, \frac{\tau}{b_j}) \right),$$
$$= *_{i,j} \left(\mu(E_i \cap F_j, \frac{s}{a_i}) * \mu((E_i \cap F_j, \frac{\iota}{b_j})) \right),$$
$$\leq *_{i,j} \left(\min \left\{ \mu(E_i \cap F_j, \frac{s}{a_i}), \mu((E_i \cap F_j, \frac{\tau}{b_j})) \right\} \right).$$

Now, we assume $\frac{s}{a_i} < \frac{\tau}{b_j}$. From (10), we conclude

$$N(\phi, s) * N(\psi, \tau) \leq *_{i,j} \mu \left(E_i \cap F_j, \frac{s}{a_i} \right). \tag{11}$$

Again, from $\dfrac{s}{a_i} < \dfrac{\tau}{b_j}$, we get $\dfrac{s}{a_i} < \dfrac{\tau+s}{a_i+b_j}$ because

$$b_j s < a_i \tau,$$

then

$$a_i s + b_j s < a_i s + a_i \tau,$$

and

$$(a_i + b_j)s < a_i(\tau + s),$$

and so

$$\dfrac{s}{a_i} < \dfrac{\tau+s}{a_i+b_j}.$$

Therefore, from (11) we have

$$N(\phi,s) * N(\psi,\tau) \leq *_{i,j}\mu\left(E_i \cap F_j, \dfrac{s}{a_i}\right), \tag{12}$$

and

$$*_{i,j}\mu\left(E_i \cap F_j, \dfrac{s}{a_i}\right) \leq *_{i,j}\mu\left(E_i \cap F_j, \dfrac{\tau+s}{a_i+b_j}\right). \tag{13}$$

From (12) and (13) we have

$$N(\phi,s) * N(\psi,\tau) \leq *_{i,j}\mu\left(E_i \cap F_j, \dfrac{\tau+s}{a_i+b_j}\right),$$
$$= N\left(\phi+\psi, s+\tau\right).$$

Now let $f, g \in L^+$, then there exist $\{\phi_n\} \subseteq L^+$ such that $\phi_n \uparrow f$. Similarly, there exist $\{\psi_n\} \subseteq L^+$ such that $\psi_n \uparrow g$, and $\phi_n + \psi_n \uparrow f + g$, then

$$\inf\left\{\int \left(\phi_n + \psi_n\right) d\mu(x, \tau+s)\right\} = \int \left(f+g\right) d\mu(x, \tau+s).$$

Also according to (12), we get

$$\int \left(\phi_n + \psi_n\right) d\mu(x, \tau+s) \geq \int \phi_n d\mu(x,s) * \int \psi_n d\mu(x,\tau),$$

and

$$\int \left(f+g\right) d\mu(x,\tau+s) = \inf\left\{\int \left(\phi_n+\psi_n\right) d\mu(x,\tau+s)\right\}$$
$$\geq \inf\left\{\int \phi_n d\mu(x,s) * \int \psi_n d\mu(x,\tau)\right\},$$
$$\geq \inf\left\{\int \phi_n d\mu(x,s)\right\} * \inf \int \psi_n d\mu(x,\tau)$$
$$= \int f d\mu(x,s) * \int g d\mu(x,\tau),$$

then
$$\int \left(f+g\right)d\mu(x,\tau+s) \geq \int fd\mu(x,s) * \int gd\mu(x,\tau).$$

(FN5) Let $f = \sum_{i=1}^{k} a_i \chi_{E_i}$, then
$$N(f,\tau_n) = \int \sum_{i=1}^{k} a_i \chi_{E_i} d\mu(x,\tau_n),$$
$$= *_{i=1}^{k} \mu\left(E_i, \frac{\tau_n}{a_i}\right),$$

and
$$\lim_{\tau_n \to \tau_0} N(f,\tau_n) = \lim *_{i=1}^{k} \mu\left(E_i, \frac{\tau_n}{a_i}\right).$$

According to Definition 5 (iii), we get
$$\lim_{\tau_n \to \tau_0} N(f,\tau_n) = \lim_{\tau_n \to \tau_0} *_{i=1}^{k} \mu\left(E_i, \frac{\tau_n}{a_i}\right),$$
$$= *_{i=1}^{k} \lim_{\tau_n \to \tau_0} \left(E_i, \frac{\tau_n}{a_i}\right),$$

and by Definition 5 (i),
$$\lim_{\tau_n \to \tau_0} N(f,\tau_n) = *_{i=1}^{k} \lim_{\tau_n \to \tau_0} \left(E_i, \frac{\tau_n}{a_i}\right),$$
$$= *_{i=1}^{k} \mu\left(E_i, \frac{\tau_0}{a_i}\right),$$
$$= \int fd\mu(x,\tau_0),$$
$$= N(f,\tau_0).$$

Now, let $f \in L^+$, then
$$N(f,\tau_n) = \int fd\mu(x,\tau_n),$$
$$= \inf\left\{\int \psi_m d\mu(x,\tau_n) | \psi_m \uparrow f\right\},$$
$$= \lim_{m \to \infty} \int \phi_m d\mu(x,\tau_n).$$

and
$$\lim_{\tau_n \to \tau_0} N(f,\tau_n) = \lim_{\tau_n \to \tau_0} \lim_{m \to \infty} \int \phi_m d\mu(x,\tau_n),$$
$$= \lim_{\tau_n \to \tau_0} \lim_{m \to \infty} \int \sum_{i=1}^{k} a_i^m \chi_{E_i^m} d\mu(x,\tau_n),$$
$$= \lim_{\tau_n \to \tau_0} \lim_{m \to \infty} *_{i=1}^{k} \mu\left(E_i^m, \frac{\tau_n}{a_i^m}\right).$$

According to Definition 5 (v), we get

$$\lim_{\tau_n \to \tau_0} N(f, \tau_n) = \lim_{\tau_n \to \tau_0} \lim_{m \to \infty} *_{i=1}^{k} \mu\left(E_i^m, \frac{\tau_n}{a_i^m}\right),$$

$$= \lim_{m \to \infty} \lim_{\tau_n \to \tau_0} *_{i=1}^{k} \mu\left(E_i^m, \frac{\tau_n}{a_i^m}\right),$$

and by Definition 5 (iii), we get

$$\lim_{\tau_n \to \tau_0} N(f, \tau_n) = \lim_{m \to \infty} \lim_{\tau_n \to \tau_0} *_{i=1}^{k} \mu\left(E_i^m, \frac{\tau_n}{a_i^m}\right),$$

$$= \lim_{m \to \infty} *_{i=1}^{k} \lim_{\tau_n \to \tau_0} \mu\left(E_i^m, \frac{\tau_n}{a_i^m}\right).$$

Using Definition 5 (i), we get

$$\lim_{\tau_n \to \tau_0} N(f, \tau_n) = \lim_{m \to \infty} *_{i=1}^{k} \lim_{\tau_n \to \tau_0} \mu\left(E_i^m, \frac{\tau_n}{a_i^m}\right),$$

$$= \lim_{m \to \infty} *_{i=1}^{k} \mu\left(E_i^m, \frac{\tau_0}{a_i^m}\right),$$

$$= \lim_{m \to \infty} \int \phi_m d\mu(x, \tau_0),$$

$$= \inf\left\{\int \phi_m d\mu(x, \tau_0)\right\},$$

$$= \int f d\mu(x, \tau_0),$$

$$= N(f, \tau_0).$$

(FN6) Let $f = \sum_{i=1}^{k} a_i \chi_{E_i}$, then

$$N(f, \tau) = \int f d\mu(x, \tau),$$

$$= \int \sum_{i=1}^{n} a_i \chi_{E_i} d\mu(x, \tau),$$

$$= *_{i=1}^{k} \mu\left(E_i, \frac{\tau}{a_i}\right).$$

and

$$\lim_{\tau \to \tau_0} N(f, \tau) = \lim_{\tau \to \tau_0} *_{i=1}^{k} \mu\left(E_i, \frac{\tau}{a_i}\right).$$

According to Definition 5 (iii), we have

$$\lim_{\tau \to 0} N(f, \tau) = \lim_{\tau \to 0} *_{i=1}^{k} \mu\left(E_i, \frac{\tau}{a_i}\right),$$

$$= *_{i=1}^{k} \lim_{\tau \to 0} \mu\left(E_i, \frac{\tau}{a_i}\right),$$

and by Definition 5 (iv),

$$\lim_{\tau \to 0} N(f, \tau) = *_{i=1}^{k} \lim_{\tau \to 0} \mu\left(E_i, \frac{\tau}{a_i}\right),$$
$$= *_{i=1}^{k} 0,$$
$$= 0.$$

Now let $f \in L^+$, so

$$N(f, \tau) = \int f d\mu(x, \tau) = \inf\left\{\int \phi_m d\mu(x, \tau)\right\},$$
$$= \lim_{m \to \infty}\left\{\int \phi_m d\mu(x, \tau)\right\},$$
$$= \lim_{m \to \infty}\left\{N(\phi_m, \tau)\right\}.$$

Then,

$$\lim_{\tau \to 0} N(f, \tau) = \lim_{\tau \to 0} \lim_{m \to \infty}\left\{N(\phi_m, \tau)\right\},$$
$$= \lim_{\tau \to 0} \lim_{m \to \infty} *_{i=1}^{k} \mu\left(E_i^m, \frac{\tau}{a_i^m}\right).$$

According to Definition 5 (v), we get

$$\lim_{\tau \to 0} N(f, \tau) = \lim_{\tau \to 0} \lim_{m \to \infty} *_{i=1}^{k} \mu\left(E_i^m, \frac{\tau}{a_i^m}\right),$$
$$= \lim_{m \to \infty} \lim_{\tau \to 0} *_{i=1}^{k} \mu\left(E_i^m, \frac{\tau}{a_i^m}\right),$$

and from Definition 5 (iii), we get

$$\lim_{\tau \to 0} N(f, \tau) = \lim_{m \to \infty} \lim_{\tau \to 0} *_{i=1}^{k} \mu\left(E_i^m, \frac{\tau}{a_i^m}\right),$$
$$= \lim_{m \to \infty} *_{i=1}^{k} \lim_{\tau \to 0} \mu\left(E_i^m, \frac{\tau}{a_i^m}\right).$$

From Definition 5 (iv), we get

$$\lim_{\tau \to 0} N(f, \tau) = \lim_{m \to \infty} *_{i=1}^{k} \lim_{\tau \to 0} \mu\left(E_i^m, \frac{\tau}{a_i^m}\right),$$
$$= \lim_{m \to \infty} *_{i=1}^{k} 0,$$
$$= 0.$$

Similarly,

$$\lim_{\tau \to \infty} N(f, \tau) = 1.$$

□

We have proved $(L^+, N, *)$ is a $*$-fuzzy normed space. Define $M : L^+ \times L^+ \times (0, \infty) \longrightarrow (0, 1]$ by

$$M(f,g,\tau) = N\left(|f-g|,\tau\right) = \int |f-g| d\mu(x,\tau),$$

then M is a fuzzy metric on L^+ and $(L^+, M, *)$ is called the $*$-fuzzy metric induced by the $*$-fuzzy normed space $(L^+, N, *)$.

Theorem 5 ([3])**.** *If $f \in L^+$ and $\varepsilon > 0$, there is an integrable fuzzy simple function $\phi = \sum_{j=1}^{n} a_j \chi_{E_j}$ such that $\int |f - \phi| d\mu(x,\tau) > 1 - \varepsilon$ for each $\tau > 0$ (that is, the integrable simple functions are dense in L^+).*

Now, we show L^+ is a complete space.

Theorem 6. *L^+ is a $*$-fuzzy Banach space.*

Proof. Let $\{f_n\} \subseteq L^+$ is a Cauchy sequence, then $\{f_n(x)\} \subset \mathbb{R}^+$ is a Cauchy sequence for every $x \in X$ and $\mathbb{R}$ is complete so there exist $y \in \mathbb{R}$ such that $f_n(x) \longrightarrow y$. We get $f : X \longrightarrow \mathbb{R}, f(x) = y$ according to corollary 3.16 [3], f is fuzzy measurable so $f \in L_+$ and according to Theorem (3), $f \in L^+$ so, $\lim_{n \to \infty} f_n(x) = f(x)$ almost everywhere or $\lim_{n \to \infty} f_n = f$. □

4. $*$-Fuzzy $(L^+)^p$ Spaces

In this section, by the concept of fuzzy measurable functions and fuzzy integrable functions we define a class of function spaces.

Definition 8. *Let $(X, \mathcal{C}, *)$ be a $*$-fuzzy measure space. We define*

$$(L^+)^p = \left\{ f : X \longrightarrow \mathbb{R}^+ \text{ in which } f \text{ is fuzzy measurable function and } \int f^p d\mu(x,\tau) > 0, p \geq 1 \right\}.$$

There is an order on $((L^+)^p, \leq)$ such that $f, g \in (L^+)^p$ we have $f \leq g$ if and only if $f(x) \leq g(x)$. Furthermore, if $f, g \in (L^+)^p$ then $|f - g| \in (L^+)^p$, and $|f - g|^p \leq f^p$ or g^p hence $\int |f - g|^p d\mu(x,\tau) \geq \max[\int f^p d\mu(x,\tau), \int g^p d\mu(x,\tau)]$.

In the next theorem we prove $*$-fuzzy $(L^+)^p$ is a $*$- fuzzy normed space.

Theorem 7. *Define $N_p : (L^+)^p \times (0,\infty) \longrightarrow (0,1]$ by $N_p(f,\tau) = \int f^p d\mu(x,\tau)$ then $((L^+)^p, N_p, *)$ is a $*$-fuzzy normed space.*

Proof.

(FN1) $N_p(f,\tau) = \int f^p d\mu(x,\tau) > 0$.
(FN2) By theorem 4.5 of [3] we have,
 $N_p(f,\tau) = 1 \iff \int f^p d\mu(x,\tau) = 1 \iff f^p = 0 \iff f = 0$, almost everywhere.
(FN3) Let $f = \phi = \sum_{i=1}^{n} a_i \chi_{E_i}$ then,

$$N_p(c\phi,\tau) = \int (c\phi)^p d\mu, \tag{14}$$

$$= \int \left(\sum_{i=1}^{n} c a_i \chi_{E_i} \right)^p d\mu,$$

$$= *_{i=1}^{n} \mu\left(E_i, \frac{\tau}{c^p a_i^p} \right).$$

On the other hand,

$$N_p(\phi, \frac{\tau}{c^p}) = \int \phi^p d\mu(x, \frac{\tau}{c^p}), \tag{15}$$

$$= \int \left(\sum_{i=1}^n a_i \chi_{E_i}\right)^p d\mu(x, \frac{\tau}{c^p}),$$

$$= \int \sum_{i=1}^n a_i^p \chi_{E_i} d\mu(x, \frac{\tau}{c^p}),$$

$$= *_{i=1}^n \mu\left(E_i, \frac{\tau}{c^p a_i^p}\right).$$

From (14) and (15) we conclude that

$$N_p(cf, \tau) = N_p\left(f, \frac{\tau}{c}\right).$$

Now let $f \in (L^+)^p$, then we have

$$N_p(cf, \tau) = \int (cf)^p d\mu(x, \tau) = \inf\left\{\int (c\phi_n)^p d\mu(x, \tau) : (c\phi_n)^p \uparrow (cf)^p\right\}. \tag{16}$$

On the other hand,

$$N_p(f, \frac{\tau}{c}) = \int f^p d\mu(x, \frac{\tau}{c}) \tag{17}$$

$$= \inf\left\{\int \phi_n^p d\mu(x, \frac{\tau}{c}) : \phi_n^p \uparrow f_n^p\right\}.$$

From (14) and (15) we get

$$\int (c\phi_n)^p d\mu(x, \tau) = N_p(c\phi_n, \tau) = N_p(\phi_n, \frac{\tau}{c}) = \int \phi_n^p d\mu(x, \frac{\tau}{c}).$$

Using (16) and (17) we get

$$N_p(cf, \tau) = N_p(f, \frac{\tau}{c}).$$

(FN4) Let $f = \phi$ and $g = \psi$ be simple functions. Then,

$$N_p\left(\phi + \psi, s + \tau\right) = N_p\left(\sum_{i=1}^n a_i \chi_{E_i} + \sum_{j=1}^m b_j \chi_{F_j}, s + \tau\right), \tag{18}$$

$$= N_p\left(\sum_{i,j}(a_i + b_j)\chi_{E_i \cap F_j}, s + \tau\right),$$

$$= \int \left(\sum_{i,j}(a_i + b_j)\chi_{E_i \cap F_j}\right)^p d\mu(x, s + \tau),$$

$$= \int \sum_{i,j}(a_i + b_j)^p \chi_{E_i \cap F_j} d\mu(x, s + \tau),$$

$$= *_{i,j} \mu\left(E_i \cap F_j, \frac{s + \tau}{(a_i + b_j)^p}\right).$$

On the other hand,

$$N_p(\phi, s) * N_p(\psi, \tau) = \left(\int \phi^p d\mu(x,s) \right) * \left(\int \psi^p d\mu(x,\tau) \right), \tag{19}$$

$$= \left(\int \left(\sum_{i=1}^n a_i \chi_{E_i \cap F_j} \right)^p d\mu(x,s) \right) * \left(\int \left(\sum_{j=1}^m b_j \chi_{E_i \cap F_j} \right)^p d\mu(x,\tau) \right),$$

$$= \left(\int \sum_{i=1}^n a_i^p \chi_{E_i \cap F_j} d\mu(x,s) \right) * \left(\int \sum_{j=1}^m b_j^p \chi_{E_i \cap F_j} d\mu(x,\tau) \right),$$

$$= \left(*_{i,j} \mu\left(E_i \cap F_j, \frac{s}{a_i^p} \right) \right) * \left(*_{i,j} \mu\left(E_i \cap F_j, \frac{\tau}{b_j^p} \right) \right),$$

$$= *_{i,j} \left(\mu\left(E_i \cap F_j, \frac{s}{a_i^p} \right) * \mu\left(E_i \cap F_j, \frac{\tau}{b_j^p} \right) \right),$$

$$\leq *_{i,j} \left(\mu\left(E_i \cap F_j, \min\left\{ \frac{s}{a_i^p}, \frac{\tau}{b_j^p} \right\} \right) \right)$$

$$\leq *_{i,j} \mu\left(E_i \cap F_j, \frac{s+\tau}{(a_i+b_j)^p} \right).$$

(FN5) Let $f = \sum_{i=1}^k a_i \chi_{E_i}$, then

$$N_p(f, \tau_n) = \int \left(\sum_{i=1}^k a_i \chi_{E_i} \right)^p d\mu(x, \tau_n),$$

$$= *_{i=1}^k \mu\left(E_i, \frac{\tau_n}{(a_i)^p} \right),$$

and so

$$\lim_{\tau_n \to \tau_0} N_p(f, \tau_n) = \lim *_{i=1}^k \mu\left(E_i, \frac{\tau_n}{(a_i)^p} \right).$$

Using Definition 5 (iii), we get

$$\lim_{\tau_n \to \tau_0} N_p(f, \tau_n) = \lim_{\tau_n \to \tau_0} *_{i=1}^k \mu\left(E_i, \frac{\tau_n}{(a_i)^p} \right)$$

$$= *_{i=1}^k \lim_{\tau_n \to \tau_0} \mu\left(E_i, \frac{\tau_n}{(a_i)^p} \right),$$

and according to Definition 5 (i),

$$\lim_{\tau_n \to \tau_0} N_p(f, \tau_n) = *_{i=1}^k \lim_{\tau_n \to \tau_0} \mu\left(E_i, \frac{\tau_n}{(a_i)^p} \right)$$

$$= *_{i=1}^k \mu\left(E_i, \frac{\tau_0}{(a_i)^p} \right)$$

$$= \int f^p d\mu(x, \tau_0),$$

$$= N_p(f, \tau_0).$$

Now let $f \in (L^+)^p$, we have

$$N_p(f, \tau_n) = \int f^p d\mu(x, \tau_n)$$
$$= \inf\left\{\int (\phi_m)^p d\mu(x, \tau_n) | \phi_m \uparrow f\right\}$$
$$= \lim_{m \to \infty} \int (\phi_m)^p d\mu(x, \tau_n).$$

Then,

$$\lim_{\tau_n \to \tau_0} N_p(f, \tau_n) = \lim_{\tau_n \to \tau_0} \lim_{m \to \infty} \int (\phi_m)^p d\mu(x, \tau_n),$$
$$= \lim_{\tau_n \to \tau_0} \lim_{m \to \infty} \int \left(\sum_{i=1}^{k} (a_i^m \chi_{E_i^m})^p d\mu(x, \tau_n)\right)$$
$$= \lim_{\tau_n \to \tau_0} \lim_{m \to \infty} *_{i=1}^{k} \mu\left(E_i^m, \frac{\tau_n}{(a_i^m)^p}\right).$$

Using Definition 5 (v), we get

$$\lim_{\tau_n \to \tau_0} N_p(f, \tau_n) = \lim_{\tau_n \to \tau_0} \lim_{m \to \infty} *_{i=1}^{k} \mu\left(E_i^m, \frac{\tau_n}{(a_i^m)^p}\right),$$
$$= \lim_{m \to \infty} \lim_{\tau_n \to \tau_0} *_{i=1}^{k} \mu\left(E_i^m, \frac{\tau_n}{(a_i^m)^p}\right),$$

and according to Definition 5 (iii)

$$\lim_{\tau_n \to \tau_0} N_p(f, \tau_n) = \lim_{m \to \infty} \lim_{\tau_n \to \tau_0} *_{i=1}^{k} \mu\left(E_i^m, \frac{\tau_n}{(a_i^m)^p}\right),$$
$$= \lim_{m \to \infty} *_{i=1}^{k} \lim_{\tau_n \to \tau_0} \mu\left(E_i^m, \frac{\tau_n}{(a_i^m)^p}\right).$$

By Definition 5 (i), we have

$$\lim_{\tau_n \to \tau_0} N_p(f, \tau_n) = \lim_{m \to \infty} *_{i=1}^{k} \lim_{\tau_n \to \tau_0} \mu\left(E_i^m, \frac{\tau_n}{(a_i^m)^p}\right),$$
$$= \lim_{m \to \infty} *_{i=1}^{k} \mu\left(E_i^m, \frac{\tau_0}{(a_i^m)^p}\right),$$
$$= \lim_{m \to \infty} \int (\phi_m)^p d\mu(x, \tau_0),$$
$$= \inf\left\{\int (\phi_m)^p d\mu(x, \tau_0)\right\},$$
$$= \int f^p d\mu(x, \tau_0),$$
$$= N_p(f, \tau_0).$$

(FN6) Let $f = \sum_{i=1}^{k} a_i \chi_{E_i}$, then

$$N_p(f,\tau) = \int f^p d\mu(x,\tau),$$
$$= \int \left(\sum_{i=1}^{k} a_i \chi_{E_i}\right)^p d\mu(x,\tau),$$
$$= *_{i=1}^{k} \mu\left(E_i, \frac{\tau}{(a_i)^p}\right),$$

and so

$$\lim_{\tau \to \tau_0} N_p(f,\tau) = \lim_{\tau \to \tau_0} *_{i=1}^{k} \mu\left(E_i, \frac{\tau}{(a_i)^p}\right).$$

Using Definition 5 (iii),

$$\lim_{\tau \to 0} N_p(f,\tau) = \lim_{\tau \to 0} *_{i=1}^{k} \mu\left(E_i, \frac{\tau}{(a_i)^p}\right),$$
$$= *_{i=1}^{k} \lim_{\tau \to 0} \mu\left(E_i, \frac{\tau}{(a_i)^p}\right)$$

and by Definition 5 (iv), we have

$$\lim_{\tau \to 0} N_p(f,\tau) = *_{i=1}^{k} \lim_{\tau \to 0} \mu\left(E_i, \frac{\tau}{(a_i)^p}\right),$$
$$= *_{i=1}^{k} 0,$$
$$= 0.$$

Now, let $f \in (L^+)^p$, then

$$N_P(f,\tau) = \int f^p d\mu(x,\tau) = \inf\left\{\int (\phi_m)^p d\mu(x,\tau) : \phi_m \uparrow f\right\},$$
$$= \lim_{m \to \infty} \left\{\int (\phi_m)^p d\mu(x,\tau)\right\},$$

and so

$$\lim_{\tau \to 0} N_p(f,\tau) = \lim_{\tau \to 0} \lim_{m \to \infty} \left\{N_p(\phi_m, \tau)\right\},$$
$$= \lim_{\tau \to 0} \lim_{m \to \infty} *_{i=1}^{k} \mu\left(E_i^m, \frac{\tau}{(a_i^m)^p}\right).$$

Using Definition 5 (v), we get

$$\lim_{\tau \to 0} N_p(f,\tau) = \lim_{\tau \to 0} \lim_{m \to \infty} *_{i=1}^{k} \mu\left(E_i^m, \frac{\tau}{(a_i^m)^p}\right),$$
$$= \lim_{m \to \infty} \lim_{\tau \to 0} *_{i=1}^{k} \mu\left(E_i^m, \frac{\tau}{(a_i^m)^p}\right),$$

and by Definition 5 (iii), we have

$$\lim_{\tau \to 0} N_p(f,\tau) = \lim_{m \to \infty} \lim_{\tau \to 0} *_{i=1}^{k} \mu\left(E_i^m, \frac{\tau}{(a_i^m)^p}\right),$$

$$= \lim_{m \to \infty} *_{i=1}^{k} \lim_{\tau \to 0} \mu\left(E_i^m, \frac{\tau}{(a_i^m)^p}\right).$$

from Definition 5 (iv), we get

$$\lim_{\tau \to 0} N_p(f,\tau) = \lim_{\tau \to 0} *_{i=1}^{k} 0,$$

$$= 0.$$

□

We proved $((L^+)^p, N_p, *)$ is a $*$-fuzzy normed space. Now, define the fuzzy set $M : (L^+)^p \times (L^+)^p \times (0,\infty) \longrightarrow (0,1]$ by

$$M(f,g,\tau) = N_p\left(|f-g|,\tau\right) = \int |f-g|^p d\mu(x,\tau).$$

Then, M is a fuzzy metric on $*$-fuzzy $(L^+)^p$ and $((L^+)^p, M, *)$ is called the $*$-fuzzy metric space induced by the $*$-fuzzy normed space $((L^+)^p, N_p, *)$. Now, we study further properties of $*$-fuzzy $(L^+)^p$.

Theorem 8. *For $1 \leq p < \infty$, the set of simple functions $g = \sum_{i=1}^{n} a_i \chi_{E_i}$ where $\mu(E_i, \tau) > 0$ for all $i \in \{1,2,...,n\}$ and for all $\tau > 0$, is dense in $*$-fuzzy $(L^+)^p$.*

Proof. Clearly simple functions $g = \sum_{i=1}^{n} a_i \chi_{E_i}$ are in $*$-fuzzy $(L^+)^p$. Let $f \in (L^+)^p$, by theorem 3.20 in [3] we can choose a sequence $\{f_n\}$ of simple functions such that $f_n \uparrow f$ almost everywhere, and so $(f - f_n)^p \downarrow 0$.

We assert $(f - f_n)^p \in L^+$ because

$$(f - f_n)^p \leq f^p,$$

and so

$$\int (f - f_n)^p d\mu(x,\tau) \geq \int f^p d\mu(x,\tau) > 0,$$

then $(f - f_n)^p \in L^+$ and $(f - f_n)^p \longrightarrow 0$. Using the fundamental convergence Theorem 3, we get

$$\lim_{n \to \infty} \int (f - f_n)^p d\mu(x,\tau) = \int 0 d\mu(x,\tau) = 1.$$

Then, $\lim_{n \to \infty} N_p(f - f_n, \tau) = 1$ i.e., $f_n \xrightarrow{N_p} f$. □

In the next theorem we prove that $*$-fuzzy $(L^+)^p$ spaces are complete.

Theorem 9. *For $1 \leq p < \infty$, $*$-fuzzy $(L^+)^p$ is a $*$-fuzzy Banach space.*

Proof. Let $\{f_n\} \subseteq (L^+)^p$ be a Cauchy sequence, then for every $x \in X$, $\{f_n(x)\} \subseteq \mathbb{R}$ is a Cauchy sequence in $\mathbb{R}$ and since $\mathbb{R}$ is complete, there exist $y \in \mathbb{R}$ such that $f_n(x) \longrightarrow y$, we define $f : X \longrightarrow \mathbb{R}$ by $f(x) = y$. Since $f_n \longrightarrow f$ almost everywhere, so $(f_n)^p \longrightarrow (f)^p$ almost everywhere, and $(f_n)^p \in L^+$

by the fundamental converge Theorem 3 we have $(f)^p \in L^+$ and $\lim \int (f_n)^p d\mu(x,\tau) = \int (f)^p d\mu(x,\tau)$, hence $f \in (L^+)^p$. □

5. Inequalities on ∗-Fuzzy $(L^+)^p$

In this section, we are ready to prove some important inequalities on ∗-fuzzy $(L^+)^p$.

Lemma 1 ([16]). *If $a \geq 0$, $b \geq 0$, and $0 < \lambda < 1$, then*

$$a^\lambda b^{1-\lambda} \leq \lambda a + (1-\lambda)b,$$

we have equality if and only if $a = b$.

Theorem 10 (Hölder's Inequality). *Suppose $1 < p < \infty$ and $\dfrac{1}{p} + \dfrac{1}{q} = 1$. If f and g are fuzzy measurable functions on X then,*

$$N(fg, \tau) \geq N_p\left(f, (p)^{\frac{1}{p}}\tau\right) * N_q\left(g, (q)^{\frac{1}{q}}\tau\right).$$

Proof. We apply Lemma 1 with $(f(x))^p = a$, $b = (g(x))^q$, and $\lambda = \dfrac{1}{p}$ to obtain

$$\left((f(x))^p\right)^{\frac{1}{p}} \cdot \left((g(x))^q\right)^{1-\frac{1}{p}} \leq \frac{1}{p}(f(x))^p + (1-\frac{1}{p})(g(x))^q,$$

then

$$f(x).g(x) \leq \left((\frac{1}{p})^{\frac{1}{p}}f(x)\right)^p + \left((\frac{1}{q})^{\frac{1}{q}}g(x)\right)^q.$$

Takeing integral of both sides, we get

$$\int f(x).g(x)d\mu(x,\tau) \geq \int \left[\left((\frac{1}{p})^{\frac{1}{p}}f(x)\right)^p + \left((\frac{1}{q})^{\frac{1}{q}}g(x)\right)^q\right]d\mu(x,\tau),$$

$$\geq \left(\int \left((\frac{1}{p})^{\frac{1}{p}}f(x)\right)^p d\mu(x,\tau)\right) * \left(\int \left((\frac{1}{q})^{\frac{1}{q}}g(x)\right)^q d\mu(x,\tau)\right),$$

$$= N_p\left((\frac{1}{p})^{\frac{1}{p}}f, \tau\right) * N_q\left((\frac{1}{q})^{\frac{1}{q}}g, \tau\right),$$

$$= N_p\left(f, (p)^{\frac{1}{p}}\tau\right) * N_q\left(g, (q)^{\frac{1}{q}}\tau\right).$$

Then,

$$N_1\left(f.g, \tau\right) \geq N_p\left(f, (p)^{\frac{1}{p}}\tau\right) * N_q\left(g, (q)^{\frac{1}{q}}\tau\right).$$

□

In the next theorem we compare two ∗-fuzzy $(L^+)^p$ spaces.

Theorem 11. *If $0 < p < q < r < \infty$, then $(L^+)^q \subseteq (L^+)^p + (L^+)^r$, that is, each $f \in (L^+)^q$ is the sum of a function in ∗-fuzzy $(L^+)^p$ and a function in ∗-fuzzy $(L^+)^r$.*

Proof. If $f \in (L^+)^q$, let $E = \{x : f(x) > 1\}$ and set $g = f\chi_E$ and $h = f\chi_{E^c}$, then

$$\begin{aligned} f &= f.1, \\ &= f(\chi_E + \chi_{E^c}), \\ &= f\chi_E + f\chi_{E^c}, \\ &= g + h. \end{aligned}$$

However,

$$g^p = (f\chi_E)^p = f^p \chi_E \leq f^q \chi_E,$$

then,

$$\int g^p d\mu \geq \int f^q \chi_E d\mu > 0,$$

then,

$$g \in (L^+)^p.$$

On the other hand,

$$h^r = (f\chi_{E^c})^r = f^r \chi_{E^c} \leq f^q \chi_{E^c},$$

then,

$$\int h^r d\mu \geq \int f^q \chi_{E^c} d\mu > 0,$$

and so

$$h \in (L^+)^r.$$

□

Now, we apply Hölder's inequality Theorem 10 to prove next theorem.

Theorem 12. *If $0 < p < q < r < \infty$, then $L^p \cap L^r \subseteq L^q$ and*

$$N_q(f, \tau) \geq N_p\left(f, \left(\frac{p}{\lambda q}\right)^{\frac{1}{p}} \tau\right) * N_r\left(f, \left(\frac{r}{(1-\lambda)q}\right)^{\frac{1}{r}} \tau\right),$$

where $\lambda \in (0,1)$ is defined by $\lambda = \dfrac{\frac{1}{q} - \frac{1}{r}}{\frac{1}{p} - \frac{1}{r}}$.

Proof. From $\int f^q d\mu(x,\tau) = \int f^{\lambda q}.f^{(1-\lambda)q} d\mu(x,\tau)$ and Hölder's inequality Theorem 10, we have

$$\int f^q d\mu(x,\tau) = \int f^{\lambda q}.f^{q(1-\lambda)} d\mu(x,\tau),$$

$$\geq \left(\int \left((\frac{\lambda q}{p})^{\frac{\lambda q}{p}} f^{\lambda q}\right)^{\frac{p}{\lambda q}} d\mu(x,\tau)\right) * \left(\int \left(\frac{(1-\lambda)q}{r}\right)^{\frac{(1-\lambda)q}{r}} f^{q(1-\lambda)} d\mu(x,\tau)\right)^{\frac{r}{(1-\lambda)q}},$$

$$\geq \left(\int \frac{\lambda q}{p} f^p d\mu(x,\tau)\right) * \left(\int \left(\frac{(1-\lambda)q}{r}\right) f^r d\mu(x,\tau)\right),$$

$$= \left(\int \left(\frac{\lambda q}{p}\right)^{\frac{1}{p}} f\right)^p d\mu(x,\tau)\right) * \left(\int \left(\left(\frac{(1-\lambda)q}{r}\right)^{\frac{1}{r}} f\right)^r d\mu(x,\tau)\right),$$

$$= N_p\left(\left(\frac{\lambda q}{p}\right)^{\frac{1}{p}} f,\tau\right) * N_r\left(\left(\frac{(1-\lambda)q}{r}\right)^{\frac{1}{r}} f,\tau\right),$$

$$= N_p\left(f,\left(\frac{p}{\lambda q}\right)^{\frac{1}{p}}\tau\right) * N_r\left(f,\left(\frac{r}{(1-\lambda)q}\right)^{\frac{1}{r}}\tau\right).$$

then,

$$N_q(f,\tau) \geq N_p\left(f,\left(\frac{p}{\lambda q}\right)^{\frac{1}{p}}\tau\right) * N_r\left(f,\left(\frac{r}{(1-\lambda)q}\right)^{\frac{1}{r}}\tau\right).$$

□

Another application of Hölder's inequality Theorem 10 helps us to prove next theorem.

Theorem 13. *If $\mu(X,\tau) > 0$ and $0 < p < q < \infty$, then $L^p(\mu) \supset L^q(\mu)$ and,*

$$N_p(f,\tau) \geq N_q\left(f,(\frac{q}{p})^{\frac{p}{q}}\tau\right) * \mu\left(X,(\frac{q}{q-p})^{\frac{q-p}{q}}\tau\right).$$

Proof. By Theorem 7 and Hölder's inequality Theorem 10, we get

$$N_p(f,\tau) = \int f^p.1 d\mu(x,\tau),$$

$$\geq N_{\frac{q}{p}}\left(f^p,(\frac{q}{p})^{\frac{p}{q}}\tau\right) * N_{\frac{q}{q-p}}\left(1,(\frac{q}{q-p})^{\frac{q-p}{q}}\tau\right),$$

$$= \int (f^p)^{\frac{q}{p}} d\mu\left(x,(\frac{q}{p})^{\frac{p}{q}}\tau\right) * \int 1 d\mu\left(x,(\frac{q}{q-p})^{\frac{q-p}{q}}\tau\right),$$

$$= \int f^q d\mu\left(x,(\frac{q}{p})^{\frac{p}{q}}\tau\right) * \mu\left(X,(\frac{q}{q-p})^{\frac{q-p}{q}}\tau\right),$$

$$= N_q\left(f,(\frac{q}{p})^{\frac{p}{q}}\tau\right) * \mu\left(X,(\frac{q}{q-p})^{\frac{q-p}{q}}\tau\right).$$

□

Finally, we prove the Chebyshev's Inequality in $*$-fuzzy $(L^+)^p$ spaces.

Theorem 14 (Chebyshev's Inequality). *If $f \in (L^+)^p(0 < p < \infty)$ then for any $a > 0$, $N_p(f,\tau) \leq N_p(\chi_{E_a}, \frac{\tau}{a})$ with respect to $E_a = \{x : f(x) > a\}$.*

Proof. We have,

$$f^p > (f\chi_{E_a})^p = f^p \chi_{E_a},$$

then

$$\int f^p d\mu(x,\tau) \leq \int f^p d\mu(x,\tau)\chi_{E_a} = \int_{E_a} f^p d\mu(x,\tau), \qquad (20)$$

and on E_a we have

$$\int_{E_a} f^p d\mu(x,\tau) \leq \int_{E_a} a^p d\mu(x,\tau) = \int a^p \chi_{E_a} d\mu(x,\tau). \qquad (21)$$

By (20) and (21) we get

$$\int f^p d\mu(x,\tau) \leq \int a^p \chi_{E_a} d\mu(x,\tau),$$
$$= \int \left(a\chi_{E_a}\right)^p d\mu(x,\tau).$$

Then,

$$N_p(f,\tau) \leq N_p(a\chi_{E_a},\tau),$$
$$= N_p(\chi_{E_a}, \frac{\tau}{a}).$$

□

6. Conclusions

We have considered an uncertainty measure μ based on the concept of fuzzy sets and continuous triangular norms named by $*$-fuzzy measure. In fact, we worked on a new model of the fuzzy measure theory ($*$-fuzzy measure) which is a dynamic generalization of the classical measure theory. $*$-fuzzy measure theory has gotten by replacing the non-negative real range and the additivity of classical measures with fuzzy sets and triangular norms. Moreover, the $*$-fuzzy measure theory has been motivated by defining new additivity property using triangular norms. Our approach can be apply for decision making problems [8,9].

We have restricted fuzzy measurable functions and fuzzy integrable functions and defined important classes of function spaces named by $*$-fuzzy $(L^+)^p$. Moreover, we have got a norm on $*$-fuzzy $(L^+)^p$ spaces and proved that $*$-fuzzy $(L^+)^p$ spaces are $*$-fuzzy Banach spaces. Finally, we have proved Chebyshev's Inequality and Hölder's Inequality.

Author Contributions: Formal analysis, A.G. and R.M.; Methodology, A.G. and R.S.; Project administration, R.M.; Resources, A.G.; Supervision, R.S.; Writing—review & editing, R.M. All authors have read and agreed to the published version of the manuscript.

Funding: The work of the third author on this paper was supported by grants APVV-18-0052 and by the project of Grant Agency of the Czech Republic (GACR) No. 18-06915S.

Acknowledgments: The authors are thankful to the anonymous referees for giving valuable comments and suggestions which helped to improve the final version of this paper.

Conflicts of Interest: The authors declare no conflict of interest.

References

1. Candeloro, D.; Mesiar, R.; Sambucini, A.R. A special class of fuzzy measures: Choquet integral and applications. *Fuzzy Sets Syst.* **2019**, *355*, 83–99.
2. Pap, E. Some elements of the classical measure theory. In *Handbook of Measure Theory*; Elsevier: Amsterdam, The Netherlands, 2002; Volumes I and II, pp. 27–82.
3. Gaffari, A.; Saadati, R.; Mesiar, R. Triangular norm-based ∗-fuzzy measure and integration. **2020**, preprint.
4. Bartwal, A.; Dimri, R.C.; Prasad, G. Some fixed point theorems in fuzzy bipolar metric spaces. *J. Nonlinear Sci. Appl.* **2020**, *13*, 196–204.
5. George, A.; Veeramani, P. On some results in fuzzy metric spaces. *Fuzzy Sets Syst.* **1994**, *64*, 395–399.
6. Gregori, V.; Mioana, J.-J.; Miravet, D. Contractive sequences in fuzzy metric spaces. *Fuzzy Sets Syst.* **2020**, *379*, 125–133.
7. Tian, J.-F.; Ha, M.-H.; Tian, D.-Z. Tripled fuzzy metric spaces and fixed point theorem. *Inform. Sci.* **2020**, *518*, 113–126.
8. Das, S.; Kar, M.B.; Kar, S.; Pal, T. An approach for decision making using intuitionistic trapezoidal fuzzy soft set. *Ann. Fuzzy Math. Inform.* **2018**, *16*, 99–116.
9. Si, A.; Das, S.; Kar, S. An approach to rank picture fuzzy numbers for decision making problems. *Decis. Mak. Appl. Manag. Eng.* **2019**, *2*, 54–64.
10. Hadzic, O.; Pap, E. *Fixed Point Theory in Probabilistic Metric Spaces*; Mathematics and its Applications; Kluwer Academic Publishers: Dordrecht, The Netherlands, 2001; Volume 536.
11. Schweizer, B.; Sklar, A. *North-Holland Series in Probability and Applied Mathematics*; North-Holland Publishing Co.: New York, NY, USA, 1983.
12. Cheng, S.C.; Mordeson, J.N. Fuzzy linear operators and fuzzy normed linear spaces. *Bull. Calcutta Math. Soc.* **1994**, *86*, 429–436.
13. Nadaban, S.; Binzar, T.; Pater, F. Some fixed point theorems for φ-contractive mappings in fuzzy normed linear spaces. *J. Nonlinear Sci. Appl.* **2017**, *10*, 5668–5676.
14. Saadati, R.; Vaezpour, S.M. Some results on fuzzy Banach spaces. *J. Appl. Math. Comput.* **2005**, *17*, 475–484.
15. Saadati, R. Nonlinear contraction and fuzzy compact operator in fuzzy Banach algebras. *Fixed Point Theory* **2019**, *20*, 289–297.
16. Folland, G.B. *Real Analysis: Modern Techniques and Their Applications*, 2nd ed.; Pure and Applied Mathematics (New York); A Wiley-Interscience Publication; John Wiley & Sons, Inc.: New York, NY, USA, 1999.

Publisher's Note: MDPI stays neutral with regard to jurisdictional claims in published maps and institutional affiliations.

© 2020 by the authors. Licensee MDPI, Basel, Switzerland. This article is an open access article distributed under the terms and conditions of the Creative Commons Attribution (CC BY) license (http://creativecommons.org/licenses/by/4.0/).

Article
Applications of Stieltjes Derivatives to Periodic Boundary Value Inclusions

Bianca Satco [1],* and George Smyrlis [2]

[1] Faculty of Electrical Engineering and Computer Science, Development and Innovation in Advanced Materials, Nanotechnologies, and Distributed Systems for Fabrication and Control (MANSiD), Integrated Center for Research, Stefan cel Mare University of Suceava, Universitatii 13, 720225 Suceava, Romania
[2] Department of Mathematics, School of Applied Mathematics and Physics, National Technical University of Athens, Zografou Campus, 157 80 Athens, Greece; gsmyrlis@math.ntua.gr
* Correspondence: bisatco@usm.ro

Received: 3 November 2020; Accepted: 27 November 2020; Published: 1 December 2020

Abstract: In the present paper, we are interested in studying first-order Stieltjes differential inclusions with periodic boundary conditions. Relying on recent results obtained by the authors in the single-valued case, the existence of regulated solutions is obtained via the multivalued Bohnenblust–Karlin fixed-point theorem and a result concerning the dependence on the data of the solution set is provided.

Keywords: periodic boundary value inclusion; Stieltjes derivative; Stieltjes integrals; Bohnenblust–Karlin fixed-point theorem; regulated function

1. Introduction

Allowing the study in a unique framework of many classical problems: ordinary differential or difference equations (in the case of an absolutely continuous measure—with respect to the Lebesgue measure—respectively of a discrete measure), impulsive differential problems (for a sum of Lebesgue measure with a discrete one), dynamic equations on time scales (see [1]) and generalized differential equations (e.g., [2,3]), it is clear why the theory of differential equations driven by measures has seen a significant growth (e.g., [1,4]).

Using a natural notion of Stieltjes derivative with respect to a non-decreasing function (c.f. [5], see also [6] or [7,8] for applications), measure-driven differential equations can be expressed, in an equivalent form, as a Stieltjes differential equation.

On the other hand, the set-valued setting covers a wider range of problems ([9–12], see also [13–15]), therefore passing from the single-valued to the multivalued case brings a real improvement.

Based on the results obtained in [4] for measure-driven differential equations with periodic boundary conditions, in the present paper we focus on nonlinear differential inclusions of the form:

$$\begin{cases} u'_g(t) + b(t)u(t) \in F(t, u(t)), \ t \in [0, T] \\ u(0) = u(T) \end{cases} \tag{1}$$

where u'_g denotes the Stieltjes derivative of the state u with respect to a left-continuous non-decreasing function $g : [0, T] \to \mathbb{R}$. This form is preferred since in many real-world problems the linear, respectively the nonlinear term has different practical meanings.

In the particular case of the identical function g, periodic differential problems have been widely considered in the literature; to mention only a few works, we refer to [16–18] for the single-valued setting and to [19,20] (without impulses) or [21,22] (allowing impulses) in the set-valued framework.

As far as the authors know, periodic differential problems driven by a non-decreasing left-continuous function g have been studied only in the single-valued case in [4].

Applying Bohnenblust–Karlin set-valued fixed-point theorem, we prove that the specified problem (1) possesses solutions and characterize the solutions as Stieltjes integrals with an appropriate Green function.

We then study the dependence of the solution set of (1) on the data; specifically, we want to estimate the perturbation of the corresponding solution set if perturbations occur in the values of b and F. Such an estimation is provided in the case where the multifunction does not depend on the state.

New results for impulsive periodic inclusions (studied, e.g., in [21,22]) can be deduced by considering as function g the sum of an absolutely continuous function with step functions. Moreover, no restrictions are imposed on the number of impulses (it can be countable, so Zeno behavior is allowed).

Having in mind that the theory of measure-driven equations is equivalent, in most situations, with the theory of dynamic equations on time scales ([1], see also [23]), our study could be used to deduce new existence and dependence on the data results for periodic dynamic inclusions on time scales (see [24,25]).

The outline of the paper is as follows. After introducing the notations and recalling some necessary known facts, in Section 3 we present an existence result for the single-valued case and then we proceed to the main results in Section 4: we prove (for the multivalued setting) an existence result and also a result on the dependence of the solution set on the data.

2. Notations and Known Facts

A regulated map $u : [0, T] \to \mathbb{R}^d$ [26] is a map with right and left limits $u(t+)$ and $u(s-)$ at every point $t \in [0, T)$ and $s \in (0, T]$. It is known that regulated functions have at most countably many discontinuities [27] and that the space $G([0, T], \mathbb{R}^d)$ of regulated functions $u : [0, T] \to \mathbb{R}^d$ is a Banach space with respect to the norm $\|u\|_C = \sup_{t \in [0,T]} \|u(t)\|$.

A collection $\mathcal{A} \subset G([0, T], \mathbb{R}^d)$ is said to be equiregulated if the following conditions hold:

- for each $t \in (0, T]$ and $\varepsilon > 0$, one can choose $\delta_{\varepsilon,t} \in (0, T]$ such that for all $u \in \mathcal{A}$

$$\|u(t') - u(t-)\| < \varepsilon, \text{ for every } t' \in (t - \delta_{\varepsilon,t}, t)$$

- for each $t \in [0, T)$ and $\varepsilon > 0$, one can choose $\delta_{\varepsilon,t} \in (0, T - t]$ such that for all $u \in \mathcal{A}$

$$\|u(t') - u(t+)\| < \varepsilon, \text{ for every } t' \in (t, t + \delta_{\varepsilon,t}).$$

Let us recall an Ascoli-type result.

Lemma 1. ([26], Corollary 2.4) *A set $\mathcal{A} \subset G([0, T], \mathbb{R}^d)$ is relatively compact if and only if it is equiregulated and pointwise bounded.*

It is not difficult to check that:

Remark 1. *A set $\mathcal{A}$ of regulated functions is equiregulated if*

$$\|u(t) - u(t')\| \leq |\chi(t) - \chi(t')|, \quad \forall\, 0 \leq t < t' \leq T, \quad \forall\, u \in \mathcal{A}$$

for some regulated function $\chi : [0, T] \to \mathbb{R}$.

In the whole paper, $g : [0, T] \to \mathbb{R}$ will be a non-decreasing left-continuous function and μ_g the Stieltjes measure defined by g. Without any loss of generality, suppose $g(0) = 0$. We deal with the Kurzweil–Stieltjes integral; we recall below the basic facts concerning this integral.

Definition 1. (Refs [2,3,27,28] or [29]) *One says that $f : [0, T] \to \mathbb{R}^d$ is Kurzweil–Stieltjes integrable (or KS-integrable) with respect to $g : [0, T] \to \mathbb{R}$ if there is $\int_0^T f(s)dg(s) \in \mathbb{R}^d$ with the property that for every $\varepsilon > 0$, one can find $\delta_\varepsilon : [0, T] \to \mathbb{R}_+$ satisfying*

$$\left\| \sum_{i=1}^k f(\xi_i)(g(t_i) - g(t_{i-1})) - \int_0^T f(s)dg(s) \right\| < \varepsilon$$

for every δ_ε-fine partition $\{([t_{i-1}, t_i], \xi_i) : i = 1, ..., k\}$ of $[0, T]$. (A partition $\{([t_{i-1}, t_i], \xi_i) : i = 1, ..., k\}$ of $[0, T]$ is δ_ε-fine iff $[t_{i-1}, t_i] \subset (\xi_i - \delta_\varepsilon(\xi_i), \xi_i + \delta_\varepsilon(\xi_i))$, for all $1 \leq i \leq k$).

The well-known Henstock–Kurzweil integral (see [30–32]) is recovered in the case where g is the identical function and $d = 1$.

In general, the Lebesgue–Stieltjes integrability with respect to g (i.e., the abstract Lebesgue integrability with respect to the Stieltjes measure μ_g) yields the Kurzweil–Stieltjes integrability with respect to g. When g is left-continuous and non-decreasing, by ([28], Theorem 6.11.3) (or ([27], Theorem 8.1)),

$$\int_0^t f(s)dg(s) = \int_{[0,t]} f(s)d\mu_g(s) - f(t)(g(t+) - g(t)) = \int_{[0,t)} f(s)d\mu_g(s), \forall t \in [0, T].$$

(Ref [29], Proposition 2.3.16) asserts that the KS-primitive $F : [0, T] \to \mathbb{R}^d$, $F(t) = \int_0^t f(s)dg(s)$ is regulated whenever g is regulated, it is left-continuous if g is left-continuous and for every $t \in [0, T)$,

$$F(t+) - F(t) = f(t)\left[g(t+) - g(t)\right].$$

Consequently, if g is continuous at some point, then F is also continuous.

To recall more properties of the primitive, we need a notion of (Stieltjes) derivative of a function with respect to another function, given in [5] (see also [33]).

Definition 2. *Let $g : [0, T] \to \mathbb{R}$ be non-decreasing and left-continuous. The derivative of $f : [0, T] \to \mathbb{R}^d$ with respect to g (or the g-derivative) at the point $t \in [0, T]$ is*

$$f'_g(t) = \lim_{t' \to t} \frac{f(t') - f(t)}{g(t') - g(t)} \quad \text{if } g \text{ is continuous at } t,$$

$$f'_g(t) = \lim_{t' \to t+} \frac{f(t') - f(t)}{g(t') - g(t)} \quad \text{if } g \text{ is discontinuous at } t,$$

if the limit exists.

The g-derivative has found interesting applications in solving real-world problems where periods of time where no activity occurs and instants with abrupt changes are both present, such as [7] or [8].

Define the following set:

$$D_g = \{t \in [0, T] : g(t+) - g(t) > 0\},$$

namely the collection of atoms of μ_g; remark that if $t \in D_g$, then

$$f'_g(t) = \frac{f(t+) - f(t)}{g(t+) - g(t)}.$$

There is a set where Definition 2 has no meaning, more precisely,

$$C_g = \{t \in [0, T] : g \text{ is constant on } (t - \varepsilon, t + \varepsilon) \text{ for some } \varepsilon > 0\}.$$

It is convenient, when working with the g-derivative, to also disregard the points of the set

$$N_g = \{u_n, v_n : n \in \mathbb{N}\} \setminus D_g,$$

where $C_g = \bigcup_{n \in \mathbb{N}} (u_n, v_n)$ is a pairwise disjoint decomposition of C_g (such a writing is possible due to the fact that C_g is open in the usual topology of the real line, see [5]).

To warrant this, take into account that $\mu_g(C_g) = \mu_g(N_g) = 0$ [5] and, when studying differential equations, the equation has to be satisfied μ_g-almost everywhere.

The connection between Stieltjes integrals and the Stieltjes derivative is given by Fundamental Theorems of Calculus ([5], Theorems 5.4, 6.2, 6.5).

Theorem 1. ([5], Theorem 6.5) *Let $f : [0, T] \to \mathbb{R}^d$ be KS-integrable with respect to the non-decreasing left-continuous function $g : [0, T] \to \mathbb{R}$. Then its primitive*

$$F(t) = \int_0^t f(s)\, dg(s), \quad t \in [0, T],$$

is g-differentiable μ_g-a.e. in [0,T] and $F'_g(t) = f(t)$, μ_g-a.e. in [0,T].

As our aim is to study a differential inclusion, we end this section with basic notions of set-valued analysis (the reader is referred to [34,35] or [36]).

Let $\mathcal{P}_{bc}(\mathbb{R}^d)$ be the space of all non-empty bounded, closed and convex subsets of $\mathbb{R}^d$ endowed with the Hausdorff–Pompeiu distance

$$D(A, A') = \max(e(A, A'), e(A', A)),$$

where the (Pompeiu-) excess of the set $A \in \mathcal{P}_{bc}(\mathbb{R}^d)$ over $A' \in \mathcal{P}_{bc}(\mathbb{R}^d)$ is given by

$$e(A, A') = \sup_{a \in A} \inf_{a' \in A'} \|a - a'\|.$$

If $A \in \mathcal{P}_{bc}(\mathbb{R}^d)$, denote by $|A| = D(A, \{0\}) = \sup_{a \in A} \|a\|$.

Let X, Y be Banach spaces and let $F : X \to \mathcal{P}(Y)$ be a multimapping. F is said to be upper semicontinuous at $u_0 \in X$ if for each $\varepsilon > 0$ there is $\delta_{\varepsilon, u_0} > 0$ such that whenever $\|u - u_0\| < \delta_{\varepsilon, u_0}$,

$$F(u) \subset F(u_0) + \varepsilon B,$$

B being the closed unit ball of Y.

Moreover, F has closed graph if for all $(u_n)_{n \in \mathbb{N}} \subset X$, $(v_n)_{n \in \mathbb{N}} \subset Y$ with

$$u_n \to u \in X, \quad v_n \to v \in Y, \quad v_n \in F(u_n), \quad n \in \mathbb{N},$$

we have $v \in F(u)$.

3. Preliminary Result—Existence Theory for the Single-Valued Problem

In this section, relying on the theory in [4], we present an existence result for the linear Stieltjes differential equation with periodic boundary conditions

$$\begin{cases} u'_g(t) + b(t)u(t) = f(t), & \mu_g \text{ −a.e. in } [0, T], \\ u(0) = u(T), \end{cases} \quad (2)$$

where $g : [0, T] \to \mathbb{R}$ is non-decreasing and left-continuous and $b : [0, T] \to \mathbb{R}$ is a μ_g-measurable function satisfying the non-resonance condition:

$$1 - b(t)\mu_g(\{t\}) \neq 0, \quad \text{for every } t \in [0, T]. \quad (3)$$

Definition 3. *A function $u : [0, T] \to \mathbb{R}^d$ is a solution of problem* (2) *if it is left-continuous and regulated, constant on the intervals where g is constant, g-differentiable μ_g-a.e. in $[0, T]$ satisfying*

$$u'_g(t) + b(t)u(t) = f(t), \; \mu_g - a.e. \text{ in } [0, T]$$

and

$$u(0) = u(T).$$

Let us remark that when $b \in L^1(\mu_g)$, the following condition is fulfilled:

$$\sum_{t \in D_g} \left| \log|1 - b(t)\mu_g(\{t\})| \right| < \infty. \quad (4)$$

Indeed, if D_g is countable, we note its elements by $\{\tilde{t}_n\}_{n \in \mathbb{N}}$ and we get

$$\sum_{n=1}^{\infty} |b(\tilde{t}_n)\mu_g(\{\tilde{t}_n\})| \leq \|b\|_{L^1(\mu_g)} < \infty$$

which implies $b(\tilde{t}_n)\mu_g(\{\tilde{t}_n\}) \to 0$ as $n \to \infty$. Then, since

$$\lim_{n \to \infty} \frac{|\log|1 - b(\tilde{t}_n)\mu_g(\{\tilde{t}_n\})||}{|b(\tilde{t}_n)\mu_g(\{\tilde{t}_n\})|} = 1, \quad (5)$$

(4) comes from the Limit Comparison Criterion for the convergence of numerical series. If D_g is finite, then (4) is trivially fulfilled.

It turns out (see [4]) that for some positive constant δ,

$$|1 - b(t)\mu_g(\{t\})| > \delta, \; \forall t \in D_g \, .$$

Moreover, $t \to |b(t)\mu_g(\{t\})|$ is bounded on $[0, T]$ since on $[0, T] \setminus D_g$ it vanishes, while on D_g we may see that is obviously bounded if D_g is finite, respectively $b(\tilde{t}_n)\mu_g(\{\tilde{t}_n\}) \to 0$ as $n \to \infty$ if D_g is countable.

To solve the problem (2), the sign of $1 - b(t)\mu_g(\{t\})$ has to be taken into account.

As in [7], if $b \in L^1_g([0,T])$, the set

$$D_g^- = \{t \in D_g : 1 - b(t)\mu_g(\{t\}) < 0\}$$

is finite since

$$\infty > \|b\|_{L^1_g} > \sum_{t \in D_g^-} b(t)\mu_g(\{t\}) > \sum_{t \in D_g^-} 1.$$

Denote by $t_1 < ... < t_k$ its elements and, for simplicity, let $t_0 = 0$ and $t_{k+1} = T$. Let

$$\alpha(t) = \begin{cases} 1, & \text{if } 0 \leq t \leq t_1 \\ (-1)^i, & \text{if } t_i < t \leq t_{i+1}, \ i = 1, ..., k \end{cases}$$

and

$$\tilde{b}(t) = \begin{cases} b(t), & \text{if } t \in [0,T] \setminus D_g \\ \dfrac{-\log|1 - b(t)\mu_g(\{t\})|}{\mu_g(\{t\})}, & \text{if } t \in D_g. \end{cases}$$

Applying Theorem 1, the following existence result can be proved:

Theorem 2. *Let* $b : [0,T] \to \mathbb{R}$ *be LS-integrable with respect to* g, *satisfying* (3) *and let* $f : [0,T] \to \mathbb{R}^d$ *be such that* $\tilde{f}(t) = \dfrac{f(t)}{1 - b(t)\mu_g(\{t\})}$ *is KS-integrable with respect to* g.
Denoting by

$$\tilde{g}(t,s) = \frac{1}{\alpha(T)e^{\int_0^T \tilde{b}(r)dg(r)} - 1} \begin{cases} \alpha(T)e^{\int_0^T \tilde{b}(r)dg(r) - \int_s^t \tilde{b}(r)dg(r)}, & \text{if } 0 \leq s \leq t \leq T \\ e^{-\int_s^t \tilde{b}(r)dg(r)}, & \text{if } 0 \leq t < s \leq T, \end{cases}$$

the function $u : [0,T] \to \mathbb{R}^d$,

$$u(t) = \frac{1}{\alpha(t)} \int_0^T \frac{\alpha(s)}{1 - b(s)\mu_g(\{s\})} \tilde{g}(t,s) f(s) dg(s),$$

is a solution of problem (2).

Proof. Obviously, the LS-integrability of $\tilde{b}$ with respect to g follows from condition (4) and the LS-integrability of b.

One can see that for all $t \in [0,T]$,

$$u(t) = \tag{6}$$

$$= \frac{1}{\alpha(T)e^{\int_0^T \tilde{b}(r)dg(r)} - 1} \left[\frac{\alpha(T)}{\alpha(t)} e^{\int_0^T \tilde{b}(r)dg(r)} e^{-\int_0^t \tilde{b}(r)dg(r)} \int_0^t \alpha(s) e^{\int_0^s \tilde{b}(r)dg(r)} \cdot \tilde{f}(s) dg(s) \right.$$

$$+ \left. \frac{1}{\alpha(t)} e^{-\int_0^t \tilde{b}(r)dg(r)} \int_t^T \alpha(s) e^{\int_0^s \tilde{b}(r)dg(r)} \cdot \tilde{f}(s) dg(s) \right].$$

Let $t \in [0,T] \setminus D_g$ be a point where the maps $\int_0^\cdot \tilde{b}(r)dg(r)$ and $\int_0^\cdot \alpha(s) e^{\int_0^s \tilde{b}(r)dg(r)} \cdot \tilde{f}(s)dg(s)$ are g-differentiable (we know that it happens μ_g-a.e.).

We notice that α is constant on a neighborhood of t, so, by the product differentiation rule (see [5], Proposition 2.2),

$$u'_g(t) = \frac{1}{\alpha(T)e^{\int_0^T \tilde{b}(r)dg(r)} - 1} \cdot$$
$$\left[\frac{\alpha(T)}{\alpha(t)} e^{\int_0^T \tilde{b}(r)dg(r)} e^{-\int_0^t \tilde{b}(r)dg(r)} (-\tilde{b}(t)) \int_0^t \alpha(s) e^{\int_0^s \tilde{b}(r)dg(r)} \cdot \tilde{f}(s) dg(s) \right.$$
$$+ \frac{\alpha(T)}{\alpha(t)} e^{\int_0^T \tilde{b}(r)dg(r)} e^{-\int_0^t \tilde{b}(r)dg(r)} \alpha(t) e^{\int_0^t \tilde{b}(r)dg(r)} \cdot \tilde{f}(t)$$
$$+ \frac{1}{\alpha(t)} e^{-\int_0^t \tilde{b}(r)dg(r)} (-\tilde{b}(t)) \int_t^T \alpha(s) e^{\int_0^s \tilde{b}(r)dg(r)} \cdot \tilde{f}(s) dg(s)$$
$$+ \left. \frac{1}{\alpha(t)} e^{-\int_0^t \tilde{b}(r)dg(r)} (-\alpha(t) e^{\int_0^t \tilde{b}(r)dg(r)} \cdot \tilde{f}(t)) \right]$$
$$= \frac{1}{\alpha(T)e^{\int_0^T \tilde{b}(r)dg(r)} - 1} \cdot \left\{ -\tilde{b}(t) \cdot \left[\frac{\alpha(T)}{\alpha(t)} \int_0^t \alpha(s) e^{\int_0^T \tilde{b}(r)dg(r) - \int_s^t \tilde{b}(r)dg(r)} \tilde{f}(s) dg(s) \right. \right.$$
$$+ \left. \left. \frac{1}{\alpha(t)} \int_t^T \alpha(s) e^{-\int_s^t \tilde{b}(r)dg(r)} \tilde{f}(s) dg(s) \right] + [\alpha(T) e^{\int_0^T \tilde{b}(r)dg(r)} - 1] \tilde{f}(t) \right\}$$
$$= -\tilde{b}(t) \frac{1}{\alpha(t)} \int_0^T \alpha(s) \tilde{g}(t,s) \tilde{f}(s) dg(s) + \tilde{f}(t)$$
$$= -\tilde{b}(t) u(t) + \tilde{f}(t) = -b(t) u(t) + f(t) \text{ (recall that } t \in [0, T] \setminus D_g).$$

When calculating the g-derivative of the exponential function, we used a chain rule ([5], Theorem 2.3) together with Theorem 1, namely:

$$\left(e^{-\int_0^t \tilde{b}(r)dg(r)} \right)'_g = e^{-\int_0^t \tilde{b}(r)dg(r)} \cdot \left(-\int_0^t \tilde{b}(r)dg(r) \right)'_g$$
$$= e^{-\int_0^t \tilde{b}(r)dg(r)} \cdot (-\tilde{b}(t)).$$

The equality $u'_g(t) = -b(t)u(t) + f(t)$ at the points in D_g can be proved exactly as in ([4], Theorem 17). □

Remark 2. *If we impose the LS-integrability with respect to g of f, then the LS-integrability (therefore, the KS-integrability) of $\frac{f(t)}{1-b(t)\mu_g(\{t\})}$ comes from the inequality*

$$\frac{1}{|1 - b(t)\mu_g(\{t\})|} \leq \max\left(1, \frac{1}{\delta}\right), \quad \forall t \in [0, T]$$

Remark 3. *The reciprocal assertion of Theorem 2 is also valid (see [4], Theorem 19). Specifically, if $b, \tilde{g}, f$ are as postulated in Theorem 2 and $u : [0, T] \to \mathbb{R}^d$ is a solution of (2), then*

$$u(t) = \frac{1}{\alpha(t)} \int_0^T \frac{\alpha(s)}{1 - b(s)\mu_g(\{s\})} \tilde{g}(t,s) f(s) dg(s), \quad t \in [0, T].$$

Remark 4. *As seen in [4], the application $(s', s'') \in [0, T] \times [0, T] \to e^{\int_{s'}^{s''} \tilde{b}(s)dg(s)}$ is regulated in both variables, therefore it is bounded. If*

$$M = \sup_{(s', s'') \in [0, T] \times [0, T]} e^{\int_{s'}^{s''} \tilde{b}(s)dg(s)}$$

then from the definition of $\tilde{g}$ it can easily be deduced that

$$|\tilde{g}(t,s)| \leq \frac{\max(M, M^2)}{\left|\alpha(T)e^{\int_0^T \tilde{b}(r)dg(r)} - 1\right|}, \quad \forall s, t \in [0, T].$$

Obviously, if

$$1 - b(t)\mu_g(t) > 0 \quad \text{for all } t \in [0, T],$$

then $\alpha(t) = 1$ for every $t \in [0, T]$, therefore the formulas and the computations become much simpler.

4. Main Results

4.1. Existence of Solutions

We aim to obtain the existence of solutions for the set-valued periodic boundary value problem (1):

$$\begin{cases} u'_g(t) + b(t)u(t) \in F(t, u(t)), & \mu_g - \text{a.e. in } [0, T], \\ u(0) = u(T). \end{cases}$$

The notion of solution adapted from the single-valued case (Definition 3) reads as follows.

Definition 4. *A function $u : [0, T] \to \mathbb{R}^d$ is a solution of problem (1) if it is left-continuous and regulated, constant on the intervals where g is constant, g-differentiable μ_g-a.e. in $[0, T]$ and*

$$u'_g(t) + b(t)u(t) = f(t)$$

with $f(t) \in F(t, u(t))$, $\mu_g - \text{a.e. in } [0, T]$.

We shall apply the following fixed-point theorem for multivalued operators.

Theorem 3. *(Bohnenblust–Karlin) Let X be a Banach space, $\mathcal{M} \subset X$ be closed and convex and the operator $A : \mathcal{M} \to \mathcal{P}(\mathcal{M})$ with closed, convex values be upper semicontinuous such that $A(\mathcal{M})$ is relatively compact. Then the operator has a fixed point.*

Theorem 4. *Let $b : [0, T] \to \mathbb{R}$ be LS-integrable with respect to g and suppose that (3) is fulfilled. Let $F : [0, T] \times \mathbb{R}^d \to \mathcal{P}_{bc}(\mathbb{R}^d)$ satisfy the following hypotheses:*

- *for every $t \in [0, T]$, $F(t, \cdot)$ is upper semicontinuous;*
- *for every $u \in \mathbb{R}^d$, $F(\cdot, u)$ is μ_g-measurable;*
- *there exists a function $\bar{\phi}$ LS-integrable with respect to g such that*

$$|F(t, u)| \leq \bar{\phi}(t)$$

for every $t \in [0, T], u \in \mathbb{R}^d$.

Then the Stieltjes differential inclusion (1) has solutions. Moreover, the solution set of (1) is $\|\cdot\|_C$-bounded.

Proof. Let X_g be the subspace of $G([0, T], \mathbb{R}^d)$ consisting of the functions being continuous on $[0, T] \setminus D_g$.

Condition (4) together with the LS-integrability with respect to g of b imply that $\tilde{b}$ has the same feature.

Following Remark 4, we note by

$$M = \sup_{(s', s'') \in [0, T] \times [0, T]} e^{\int_{s'}^{s''} \tilde{b}(s) dg(s)}.$$

By condition (4), for every $t \in [0, T]$,
$$\frac{1}{|1 - b(t)\mu_g(\{t\})|}|F(t,u)| \leq \max\left(1, \frac{1}{\delta}\right) \cdot \overline{\phi}(t)$$

so we shall denote by
$$\phi(t) = \max\left(1, \frac{1}{\delta}\right) \cdot \overline{\phi}(t), \; \forall t \in [0, T].$$

Consider
$$\mathcal{M} = \left\{ v \in X_g : \|v\|_C \leq \frac{\max(M, M^2)}{\left|\alpha(T)e^{\int_0^T \tilde{b}(r)dg(r)} - 1\right|} \int_0^T \phi(s)dg(s) \right\}.$$

and the operator $A : \mathcal{M} \to \mathcal{P}(\mathcal{M})$ given, for each $u \in \mathcal{M}$, by
$$Au = \left\{ v \in X_g : v(t) = \frac{1}{\alpha(t)} \int_0^T \frac{\alpha(s)}{1 - b(s)\mu_g(\{s\})} \tilde{g}(t,s)f(s)dg(s) : f \in S_{F(\cdot, u(\cdot))} \right\}$$

with $\tilde{g}$ as in Theorem 2 and
$$S_{F(\cdot, u(\cdot))} = \left\{ f \in L^1(\mu_g, \mathbb{R}^d) : f(t) \in F(t, u(t)) \; \mu_g - \text{a.e.} \right\}.$$

A is well defined: for each $u \in X_g$, $S_{F(\cdot, u(\cdot))}$ is non-empty and whenever $u \in X_g$, i.e., u is regulated and continuous on $[0,T] \setminus D_g$, each element of Au has the same feature. Indeed, we note that α is constant in a neighborhood of $t \in [0,T] \setminus D_g$, and writing each element of Au as in (6), by ([29], Proposition 2.3.16) we deduce that it is regulated and continuous on $[0,T] \setminus D_g$.

We next show that $\|u\|_C \leq \frac{\max(M,M^2)}{\left|\alpha(T)e^{\int_0^T \tilde{b}(r)dg(r)} - 1\right|} \int_0^T \phi(s)dg(s)$ implies that every $v \in Au$ satisfies the same inequality.

Indeed, fix $t \in [0, T]$. Then every $v \in Au$ is given (by the definition of the operator A) by some selection f of $F(\cdot, u(\cdot))$ and we can see, by Remark 4, that

$$\begin{aligned}
\|v(t)\| &\leq \max\left(1, \frac{1}{\delta}\right) \int_0^T |\tilde{g}(t,s)| \cdot |f(s)| \, dg(s) \\
&\leq \max\left(1, \frac{1}{\delta}\right) \frac{\max(M, M^2)}{\left|\alpha(T)e^{\int_0^T \tilde{b}(r)dg(r)} - 1\right|} \int_0^T |f(s)| dg(s) \\
&\leq \max\left(1, \frac{1}{\delta}\right) \frac{\max(M, M^2)}{\left|\alpha(T)e^{\int_0^T \tilde{b}(r)dg(r)} - 1\right|} \int_0^T \overline{\phi}(s)dg(s),
\end{aligned}$$

whence
$$\|v\|_C \leq \frac{\max(M, M^2)}{\left|\alpha(T)e^{\int_0^T \tilde{b}(r)dg(r)} - 1\right|} \int_0^T \phi(s)dg(s).$$

Let us next check that the operator has closed, convex values.

Let $u \in \mathcal{M}$. Obviously, $S_{F(\cdot, u(\cdot))}$ is convex (recall that F has convex values), therefore, Au is convex as well.

To prove that it is closed, take $(v_n)_{n \in \mathbb{N}} \subset Au$ uniformly convergent to $v \in \mathcal{M}$; specifically, for each $n \in \mathbb{N}$, one can find $f_n \in S_{F(\cdot, u(\cdot))}$ such that

$$v_n(t) = \frac{1}{\alpha(t)} \int_0^T \frac{\alpha(s)}{1 - b(s)\mu_g(\{s\})} \tilde{g}(t,s) f_n(s) dg(s), \; t \in [0,T] \text{ and } v_n \to v \text{ uniformly.}$$

One can see that
$$\|f_n(t)\| \leq \overline{\phi}(t), \ \forall n \in \mathbb{N}, \ t \in [0,T],$$
so there exists a subsequence $(f_{n_k})_{k \in \mathbb{N}}$ weakly $L^1(\mu_g, \mathbb{R}^d)$ convergent to a function $f \in L^1(\mu_g, \mathbb{R}^d)$ (Dunford-Pettis Theorem). In a classical way (Mazur's theorem and properties of norm-convergent sequences in $L^1(\mu_g, \mathbb{R}^d)$), a sequence of convex combinations tends pointwise μ_g -a.e. to f, whence
$$f(\cdot) \in S_{F(\cdot, u(\cdot))}.$$

By a dominated convergence result (see [28], Theorem 6.8.7) applied for the components of $(f_{n_k})_{k \in \mathbb{N}}$, f, one deduces that

$$v_{n_k}(t) = \frac{1}{\alpha(t)} \int_0^T \frac{\alpha(s)}{1 - b(s)\mu_g(\{s\})} \tilde{g}(t,s) f_{n_k}(s) dg(s) \rightarrow$$
$$\frac{1}{\alpha(t)} \int_0^T \frac{\alpha(s)}{1 - b(s)\mu_g(\{s\})} \tilde{g}(t,s) f(s) dg(s)$$

and so,
$$v(t) = \frac{1}{\alpha(t)} \int_0^T \frac{\alpha(s)}{1 - b(s)\mu_g(\{s\})} \tilde{g}(t,s) f(s) dg(s), \ t \in [0,T],$$

thus Au is closed.

We will prove that A satisfies the hypotheses of Theorem 3.
We check that $A(\mathcal{M})$ is relatively compact, using Lemma 1.
Take $0 \leq t < t' \leq T$.
For each $u \in \mathcal{M}$ and each $v \in Au$ (defined via a selection f of $F(\cdot, u(\cdot))$ LS-integrable with respect to g),

$$\|v(t) - v(t')\| \leq \left\| \frac{1}{\alpha(t)} \int_0^T \frac{\alpha(s)}{1 - b(s)\mu_g(\{s\})} (\tilde{g}(t,s) - \tilde{g}(t',s)) f(s)) dg(s) \right\|$$
$$+ \left\| \left(\frac{1}{\alpha(t)} - \frac{1}{\alpha(t')} \right) \int_0^T \frac{\alpha(s)}{1 - b(s)\mu_g(\{s\})} \tilde{g}(t',s) f(s)) dg(s) \right\|.$$

We note that $|\alpha(t)| = 1$ for each $t \in [0, T]$, so we can write

$$\left\| \frac{1}{\alpha(t)} \int_0^T \frac{\alpha(s)}{1 - b(s)\mu_g(\{s\})} (\tilde{g}(t,s) - \tilde{g}(t',s)) f(s) dg(s) \right\|$$

$$\leq \frac{1}{\left| \alpha(T) e^{\int_0^T \tilde{b}(r) dg(r)} - 1 \right|}$$

$$\left[\left\| \alpha(T) \int_0^t \frac{\alpha(s)}{1 - b(s)\mu_g(\{s\})} \left(e^{\int_0^T \tilde{b}(r)dg(r) - \int_s^t \tilde{b}(r)dg(r)} - e^{\int_0^T \tilde{b}(r)dg(r) - \int_s^{t'} \tilde{b}(r)dg(r)} \right) f(s) dg(s) \right\| \right.$$

$$+ \left\| \int_{t'}^T \frac{\alpha(s)}{1 - b(s)\mu_g(\{s\})} \left(e^{-\int_s^t \tilde{b}(r)dg(r)} - e^{-\int_s^{t'} \tilde{b}(r)dg(r)} \right) f(s) dg(s) \right\|$$

$$+ \left\| \int_t^{t'} \frac{\alpha(s)}{1 - b(s)\mu_g(\{s\})} \left(e^{-\int_s^t \tilde{b}(r)dg(r)} - \alpha(T) e^{\int_0^T \tilde{b}(r)dg(r) - \int_s^{t'} \tilde{b}(r)dg(r)} \right) f(s) dg(s) \right\| \right]$$

$$= \frac{1}{\left| \alpha(T) e^{\int_0^T \tilde{b}(r) dg(r)} - 1 \right|}$$

$$\left[\left\| \int_0^t \frac{\alpha(s)}{1 - b(s)\mu_g(\{s\})} e^{\int_0^T \tilde{b}(r)dg(r) - \int_s^t \tilde{b}(r)dg(r)} \left(1 - e^{-\int_t^{t'} \tilde{b}(r)dg(r)} \right) f(s) dg(s) \right\| \right.$$

$$+ \left\| \int_{t'}^T \frac{\alpha(s)}{1 - b(s)\mu_g(\{s\})} e^{-\int_s^{t'} \tilde{b}(r)dg(r)} \left(e^{\int_t^{t'} \tilde{b}(r)dg(r)} - 1 \right) f(s) dg(s) \right\|$$

$$+ \left\| \int_t^{t'} \frac{\alpha(s)}{1 - b(s)\mu_g(\{s\})} \left(e^{-\int_s^t \tilde{b}(r)dg(r)} - \alpha(T) e^{\int_0^T \tilde{b}(r)dg(r) - \int_s^{t'} \tilde{b}(r)dg(r)} \right) f(s) dg(s) \right\| \right].$$

On the other hand, again by $|\alpha(t)| = |\alpha(t')| = 1$,

$$\left\| \left(\frac{1}{\alpha(t)} - \frac{1}{\alpha(t')} \right) \int_0^T \frac{\alpha(s)}{1 - b(s)\mu_g(\{s\})} \tilde{g}(t',s) f(s) dg(s) \right\|$$

$$= |\alpha(t) - \alpha(t')| \left\| \int_0^T \frac{\alpha(s)}{1 - b(s)\mu_g(\{s\})} \tilde{g}(t',s) f(s) dg(s) \right\|$$

and using the definition of $\tilde{g}$ together with Remark 2 one gets

$$\left\| \left(\frac{1}{\alpha(t)} - \frac{1}{\alpha(t')} \right) \int_0^T \frac{\alpha(s)}{1 - b(s)\mu_g(\{s\})} \tilde{g}(t',s) f(s) dg(s) \right\|$$

$$\leq \frac{|\alpha(t) - \alpha(t')|}{\left| \alpha(T) e^{\int_0^T \tilde{b}(r) dg(r)} - 1 \right|} \left[\left\| \int_0^{t'} \frac{\alpha(s)}{1 - b(s)\mu_g(\{s\})} e^{\int_0^T \tilde{b}(r)dg(r) - \int_s^{t'} \tilde{b}(r)dg(r)} f(s) dg(s) \right\| \right.$$

$$+ \left\| \int_{t'}^T \frac{\alpha(s)}{1 - b(s)\mu_g(\{s\})} e^{-\int_s^{t'} \tilde{b}(r)dg(r)} f(s) dg(s) \right\| \right]$$

$$\leq \max\left(1, \frac{1}{\delta}\right) \frac{|\alpha(t) - \alpha(t')|}{\left| \alpha(T) e^{\int_0^T \tilde{b}(r) dg(r)} - 1 \right|} \left[\int_0^{t'} \left\| e^{\int_0^T \tilde{b}(r)dg(r) - \int_s^{t'} \tilde{b}(r)dg(r)} f(s) \right\| dg(s) \right.$$

$$+ \left. \int_{t'}^T \left\| e^{-\int_s^{t'} \tilde{b}(r)dg(r)} f(s) \right\| dg(s) \right].$$

But

$$e^{\int_0^T \tilde{b}(r)dg(r) - \int_s^t \tilde{b}(r)dg(r)} \leq M^2 \quad \text{and} \quad e^{\int_0^T \tilde{b}(r)dg(r) - \int_s^{t'} \tilde{b}(r)dg(r)} \leq M^2.$$

We thus get

$$\|v(t) - v(t')\|$$
$$\leq \frac{M}{\left|\alpha(T)e^{\int_0^T \tilde{b}(r)dg(r)} - 1\right|} \max\left(1, \frac{1}{\delta}\right) \left[M \int_0^t \left|1 - e^{-\int_t^{t'} \tilde{b}(r)dg(r)}\right| \cdot \|f(s)\| dg(s)\right.$$
$$+ \int_{t'}^T \left|e^{\int_t^{t'} \tilde{b}(r)dg(r)} - 1\right| \cdot \|f(s)\| dg(s)$$
$$+ (1+M) \int_t^{t'} \|f(s)\| dg(s)\right]$$
$$+ \frac{|\alpha(t) - \alpha(t')|}{|\alpha(T)e^{\int_0^T \tilde{b}(r)dg(r)} - 1|} \max\left(1, \frac{1}{\delta}\right) \left[M^2 \int_0^{t'} \|f(s)\| dg(s)\right.$$
$$\left. + M \int_{t'}^T \|f(s)\| dg(s)\right],$$

so, taking into account the definition of ϕ, it follows that

$$\|v(t) - v(t')\|$$
$$\leq \frac{M}{\left|\alpha(T)e^{\int_0^T \tilde{b}(r)dg(r)} - 1\right|} \left[M \left|1 - e^{-\int_t^{t'} \tilde{b}(r)dg(r)}\right| \cdot \int_0^T \phi(s) dg(s)\right.$$
$$+ \left|e^{\int_t^{t'} \tilde{b}(r)dg(r)} - 1\right| \cdot \int_0^T \phi(s) dg(s) + (1+M) \int_t^{t'} \phi(s) dg(s)$$
$$\left. + (M+1)|\alpha(t) - \alpha(t')| \int_0^T \phi(s) dg(s)\right].$$

But

$$\left|1 - e^{-\int_t^{t'} \tilde{b}(r)dg(r)}\right| \leq M \left|e^{-\int_0^t \tilde{b}(r)dg(r)} - e^{-\int_0^{t'} \tilde{b}(r)dg(r)}\right|$$

and similarly for $\left|e^{\int_t^{t'} \tilde{b}(r)dg(r)} - 1\right|$, while

$$\int_t^{t'} \phi(s) dg(s) = \int_0^{t'} \phi(s) dg(s) - \int_0^t \phi(s) dg(s).$$

Remark 1 yields now that the set $A(\mathcal{M})$ is equiregulated.

The pointwise boundedness is immediate, therefore Lemma 1 implies that $\{Au : u \in \mathcal{M}\}$ is relatively compact.

Next, let us prove that A is upper semicontinuous. As $A(\mathcal{M})$ is relatively compact, it suffices to verify that A has closed graph (see [36], Proposition 2.23).

Let $(u_n)_{n \in \mathbb{N}} \subset \mathcal{M}$ converge uniformly to $u \in \mathcal{M}$ and $(v_n)_{n \in \mathbb{N}} \subset \mathcal{M}$ converge uniformly to $v \in \mathcal{M}$ be such that $v_n \in Au_n$ for all $n \in \mathbb{N}$.

One can find, for every $n \in \mathbb{N}$, $f_n \in S_{F(\cdot, u_n(\cdot))}$ such that

$$v_n(t) = \frac{1}{\alpha(t)} \int_0^T \frac{\alpha(s)}{1 - b(s)\mu_g(\{s\})} \tilde{g}(t,s) f_n(s) dg(s), \ t \in [0, T].$$

As before,

$$\|f_n(t)\| \leq \overline{\phi}(t), \ \forall n \in \mathbb{N}, \ t \in [0, T],$$

so there is a subsequence $(f_{n_k})_{k\in\mathbb{N}}$ convergent in the weak-$L^1(\mu_g, \mathbb{R}^d)$ topology to a function $f \in L^1(\mu_g, \mathbb{R}^d)$. It follows that a sequence of convex combinations of $\{f_{n_k} : k \in \mathbb{N}\}$ tends pointwise (μ_g-a.e.) to f. On the other hand, F is upper semicontinuous with respect to the second value also with closed values, thus it has closed graph with respect to the second value (see [36], Proposition 2.17). Combining these two facts, we may easily check that

$$f \in S_{F(\cdot, u(\cdot))}.$$

By a dominated convergence result (see [28], Theorem 6.8.7) applied for the components of $(f_{n_k})_{k\in\mathbb{N}}$ and f, one deduces that the corresponding sequence of convex combinations of $(v_{n_k})_k$ converges to

$$\frac{1}{\alpha(t)} \int_0^T \frac{\alpha(s)}{1-b(s)\mu_g(\{s\})} \tilde{g}(t,s) f(s) dg(s)$$

whence

$$v(t) = \frac{1}{\alpha(t)} \int_0^T \frac{\alpha(s)}{1-b(s)\mu_g(\{s\})} \tilde{g}(t,s) f(s) dg(s), \ t \in [0, T]$$

and consequently $v \in Au$.

Finally, Bohnenblust–Karlin fixed-point theorem yields that the operator has fixed points, which are solutions to problem (1) by Theorem 2. □

4.2. Dependence on the Data

Let us now study in which manner the solution set of problem (1) depends on the data. For this purpose, we are forced to drop the dependence on the state of the right-hand side. To be more precise, if we consider functions b_1, b_2 as in Theorem 4 and multifunctions $F_1, F_2 : [0, T] \to \mathcal{P}_{bc}(\mathbb{R}^d)$ such that the considered problem has solutions, we are interested in finding the relation between the solution set $\mathcal{S}_1$ of

$$\begin{cases} u'_g(t) + b_1(t)u(t) \in F_1(t), \ \mu_g\text{-a.e. in } [0, T] \\ u(0) = u(T) \end{cases}$$

and the solution set $\mathcal{S}_2$ of

$$\begin{cases} u'_g(t) + b_2(t)u(t) \in F_2(t), \ \mu_g\text{-a.e. in } [0, T] \\ u(0) = u(T). \end{cases}$$

The perturbation of b shall be measured through

$$\|b_1 - b_2\|_C = \sup_{t \in [0,T]} |b_1(t) - b_2(t)| \quad \text{or} \quad \|b_1 - b_2\|_{L^1} = \int_0^T |b_1(s) - b_2(s)| dg(s)$$

while the perturbation of F through

$$D_C(F_1, F_2) = \sup_{t \in [0,T]} D(F_1(t), F_2(t)) \quad \text{or} \quad D_{L^1}(F_1, F_2) = \int_0^T D(F_1(s), F_2(s)) dg(s).$$

Correspondingly, one can measure the distance between the $\|\cdot\|_C$-bounded sets $\mathcal{S}_1, \mathcal{S}_2$ of regulated functions in the following ways:

$$D_C(\mathcal{S}_1, \mathcal{S}_2) = \max(e_C(\mathcal{S}_1, \mathcal{S}_2), e_C(\mathcal{S}_2, \mathcal{S}_1)),$$

where the Pompeiu-excess of the set $\mathcal{S}_1$ over the set $\mathcal{S}_2$ is defined by

$$e_C(\mathcal{S}_1, \mathcal{S}_2) = \sup_{u \in \mathcal{S}_1} \inf_{u' \in \mathcal{S}_2} \|u - u'\|_C$$

or

$$D_{L^1}(\mathcal{S}_1, \mathcal{S}_2) = \max(e_{L^1}(\mathcal{S}_1, \mathcal{S}_2), e_{L^1}(\mathcal{S}_2, \mathcal{S}_1)),$$

where the excess of $\mathcal{S}_1$ over $\mathcal{S}_2$ is

$$e_{L^1}(\mathcal{S}_1, \mathcal{S}_2) = \sup_{u \in \mathcal{S}_1} \inf_{u' \in \mathcal{S}_2} \|u - u'\|_{L^1}.$$

Let us note that

Remark 5. *For any $x_1, x_2 \in [a,b] \subset \mathbb{R}$,*

(i)
$$|e^{x_1} - e^{x_2}| \leq e^b |x_1 - x_2|.$$

(ii) *if $a > 0$,*
$$|\log|x_1| - \log|x_2|| \leq \frac{1}{a}|x_1 - x_2|.$$

Theorem 5. *Let $b_1, b_2 : [0, T] \to \mathbb{R}$ be LS-integrable with respect to g and suppose that (3) is fulfilled for both b_1 and b_2.*
Let $F_1, F_2 : [0, T] \to \mathcal{P}_{bc}(\mathbb{R}^d)$ satisfy the following hypotheses:

- *F_1, F_2 are μ_g-measurable;*
- *there exists a function $\overline{\phi}$ LS-integrable with respect to g such that*

$$|F_i(t)| \leq \overline{\phi}(t), \ i = 1, 2, \ \forall t \in [0, T].$$

Then there exist positive constants $C_i, i = \overline{1,6}$ such that for every $u_1 \in \mathcal{S}_1$, one can find $u_2 \in \mathcal{S}_2$ satisfying, for all $t \in [0, T]$,

$$\|u_1(t) - u_2(t)\|$$
$$\leq C_1 \int_0^T |b_1(s) - b_2(s)| dg(s) + C_2 |\alpha_1(T) - \alpha_2(T)|$$
$$+ C_3 \int_0^T \overline{\phi}(s)|\alpha_1(s) - \alpha_2(s)| dg(s) + C_4 \int_0^T |b_1(s) - b_2(s)|\overline{\phi}(s) dg(s)$$
$$+ C_5 \int_0^T D(F_1(s), F_2(s)) dg(s) + C_6 |\alpha_1(t) - \alpha_2(t)|.$$

Proof. Let $u_1 \in \mathcal{S}_1$. Then there exists a selection f_1 of F_1 which is LS-integrable with respect to g such that

$$u_1(t) = \frac{1}{\alpha_1(t)} \int_0^T \frac{\alpha_1(s)}{1 - b_1(s)\mu_g(\{s\})} \tilde{g}_1(t,s) f_1(s) dg(s), \ t \in [0, T],$$

where

$$D_g^{1,-} = \{t \in D_g : 1 - b_1(t)\mu_g(\{t\}) < 0\} = \{t_1^1, ..., t_k^1\}$$

(with the obvious convention $t_0^1 = 0$ and $t_{k+1}^1 = T$),

$$\alpha_1(t) = \begin{cases} 1, \ if \ 0 \leq t \leq t_1^1 \\ (-1)^i, \ if \ t_i^1 < t \leq t_{i+1}^1, \ i = 1, ..., k, \end{cases}$$

$$\tilde{b}_1(t) = \begin{cases} b_1(t), \text{ if } t \in [0,T] \setminus D_g \\ \frac{-\log|1-b_1(t)\mu_g(\{t\})|}{\mu_g(\{t\})}, \text{ if } t \in D_g \end{cases}$$

and

$$\tilde{g}_1(t,s) = \frac{1}{\alpha_1(T)e^{\int_0^T \tilde{b}_1(r)dg(r)} - 1} \begin{cases} \alpha_1(T)e^{\int_0^T \tilde{b}_1(r)dg(r) - \int_s^t \tilde{b}_1(r)dg(r)}, \text{ if } 0 \leq s \leq t \leq T \\ e^{-\int_s^t \tilde{b}_1(r)dg(r)}, \text{ if } 0 \leq t < s \leq T. \end{cases}$$

By ([34], Corollary 8.2.13) we can choose f_2 as the μ_g-measurable selection of F_2 satisfying

$$\|f_1(t) - f_2(t)\| = d(f_1(t), F_2(t)), \forall t \in [0,T]$$

so, by the very definition of the Pompeiu-Hausdorff distance,

$$\|f_1(t) - f_2(t)\| \leq D(F_1(t), F_2(t)), \forall t \in [0,T].$$

Consider now the function $u_2 : [0,T] \to \mathbb{R}^d$ given by

$$u_2(t) = \frac{1}{\alpha_2(t)} \int_0^T \frac{\alpha_2(s)}{1-b_2(s)\mu_g(\{s\})} \tilde{g}_2(t,s) f_2(s) dg(s), t \in [0,T],$$

where

$$D_g^{2,-} = \{t \in D_g : 1 - b_2(t)\mu_g(\{t\}) < 0\} = \{t_1^2, ..., t_l^2\}$$

(again, with the convention $t_0^2 = 0$ and $t_{l+1}^2 = T$),

$$\alpha_2(t) = \begin{cases} 1, \text{ if } 0 \leq t \leq t_1^2 \\ (-1)^i, \text{ if } t_i^2 < t \leq t_{i+1}^2, i = 1, ..., l, \end{cases}$$

$$\tilde{b}_2(t) = \begin{cases} b_2(t), \text{ if } t \in [0,T] \setminus D_g \\ \frac{-\log|1-b_2(t)\mu_g(\{t\})|}{\mu_g(\{t\})}, \text{ if } t \in D_g \end{cases}$$

and

$$\tilde{g}_2(t,s) = \frac{1}{\alpha_2(T)e^{\int_0^T \tilde{b}_2(r)dg(r)} - 1} \begin{cases} \alpha_2(T)e^{\int_0^T \tilde{b}_2(r)dg(r) - \int_s^t \tilde{b}_2(r)dg(r)}, \text{ if } 0 \leq s \leq t \leq T \\ e^{-\int_s^t \tilde{b}_2(r)dg(r)}, \text{ if } 0 \leq t < s \leq T. \end{cases}$$

Obviously, $u_2 \in S_2$ by Theorem 2. Let us see that it satisfies the requested inequality for some well-chosen constants $C_i, i = 1, ..., 6$.

First, we may write

$$\|u_1(t) - u_2(t)\|$$
$$= \left\| \frac{1}{\alpha_1(t)} \int_0^T \frac{\alpha_1(s)}{1-b_1(s)\mu_g(\{s\})} \tilde{g}_1(t,s) f_1(s) dg(s) \right.$$
$$\left. - \frac{1}{\alpha_2(t)} \int_0^T \frac{\alpha_2(s)}{1-b_2(s)\mu_g(\{s\})} \tilde{g}_2(t,s) f_2(s) dg(s) \right\|$$
$$\leq \left\| \frac{1}{\alpha_1(t)} \int_0^T \frac{\alpha_1(s)}{1-b_1(s)\mu_g(\{s\})} \tilde{g}_1(t,s) f_1(s) dg(s) \right.$$
$$\left. - \frac{1}{\alpha_1(t)} \int_0^T \frac{\alpha_2(s)}{1-b_2(s)\mu_g(\{s\})} \tilde{g}_2(t,s) f_2(s) dg(s) \right\|$$
$$+ \left\| \frac{1}{\alpha_1(t)} \int_0^T \frac{\alpha_2(s)}{1-b_2(s)\mu_g(\{s\})} \tilde{g}_2(t,s) f_2(s) dg(s) \right.$$
$$\left. - \frac{1}{\alpha_2(t)} \int_0^T \frac{\alpha_2(s)}{1-b_2(s)\mu_g(\{s\})} \tilde{g}_2(t,s) f_2(s) dg(s) \right\|$$

Using the remark that $|\alpha_1(t)| = 1$ and also $|\alpha_2(t)| = 1$ for every $t \in [0, T]$, we obtain

$$\|u_1(t) - u_2(t)\|$$
$$\leq \left\| \int_0^T \left(\frac{\alpha_1(s)}{1 - b_1(s)\mu_g(\{s\})} \tilde{g}_1(t,s) f_1(s) - \frac{\alpha_2(s)}{1 - b_2(s)\mu_g(\{s\})} \tilde{g}_2(t,s) f_2(s) \right) dg(s) \right\|$$
$$+ \left\| (\alpha_1(t) - \alpha_2(t)) \int_0^T \frac{\alpha_2(s)}{1 - b_2(s)\mu_g(\{s\})} \tilde{g}_2(t,s) f_2(s) dg(s) \right\| \tag{7}$$

(please note that $\left| \frac{1}{\alpha_1(t)} - \frac{1}{\alpha_2(t)} \right| = |\alpha_1(t) - \alpha_2(t)|$, for every $t \in [0, T]$).

Let us evaluate the first term of the sum (7):

$$\left\| \int_0^T \left(\frac{\alpha_1(s)}{1 - b_1(s)\mu_g(\{s\})} \tilde{g}_1(t,s) f_1(s) - \frac{\alpha_2(s)}{1 - b_2(s)\mu_g(\{s\})} \tilde{g}_2(t,s) f_2(s) \right) dg(s) \right\|$$
$$\leq \left\| \int_0^T \left(\frac{\alpha_1(s)}{1 - b_1(s)\mu_g(\{s\})} \tilde{g}_1(t,s) - \frac{\alpha_2(s)}{1 - b_2(s)\mu_g(\{s\})} \tilde{g}_2(t,s) \right) f_1(s) dg(s) \right\|$$
$$+ \left\| \int_0^T \frac{\alpha_2(s)}{1 - b_2(s)\mu_g(\{s\})} \tilde{g}_2(t,s) (f_1(s) - f_2(s)) dg(s) \right\|$$

and by Remark 2,

$$\left| \frac{\alpha_1(s)}{1 - b_1(s)\mu_g(\{s\})} \tilde{g}_1(t,s) - \frac{\alpha_2(s)}{1 - b_2(s)\mu_g(\{s\})} \tilde{g}_2(t,s) \right|$$
$$\leq \left| \frac{\alpha_1(s)}{1 - b_1(s)\mu_g(\{s\})} (\tilde{g}_1(t,s) - \tilde{g}_2(t,s)) \right| + \left| \left(\frac{\alpha_1(s)}{1 - b_1(s)\mu_g(\{s\})} - \frac{\alpha_2(s)}{1 - b_2(s)\mu_g(\{s\})} \right) \tilde{g}_2(t,s) \right|$$
$$= \left| \frac{\alpha_1(s)}{1 - b_1(s)\mu_g(\{s\})} (\tilde{g}_1(t,s) - \tilde{g}_2(t,s)) \right| + \left| \frac{\alpha_1(s) - \alpha_2(s) + (\alpha_2(s)b_1(s) - \alpha_1(s)b_2(s))\mu_g(\{s\})}{(1 - b_1(s)\mu_g(\{s\}))(1 - b_2(s)\mu_g(\{s\}))} \tilde{g}_2(t,s) \right|$$
$$\leq \max\left(1, \frac{1}{\delta_1}\right) |\tilde{g}_1(t,s) - \tilde{g}_2(t,s)|$$
$$+ \max\left(1, \frac{1}{\delta_1}\right) \max\left(1, \frac{1}{\delta_2}\right) |\tilde{g}_2(t,s)| (|\alpha_1(s) - \alpha_2(s)|$$
$$+ |\alpha_1(s) - \alpha_2(s)| \cdot |b_2(s)\mu_g(\{s\})| + |b_1(s) - b_2(s)|\mu_g(\{s\})),$$

where δ_1, δ_2 are the corresponding positive constants in Remark 2 for b_1, b_2 respectively.

Since the condition (4) is verified, $|b_2(s)\mu_g(\{s\})|$ is bounded, say by m_2. Then

$$\left\| \int_0^T \left(\frac{\alpha_1(s)}{1 - b_1(s)\mu_g(\{s\})} \tilde{g}_1(t,s) - \frac{\alpha_2(s)}{1 - b_2(s)\mu_g(\{s\})} \tilde{g}_2(t,s) \right) f_1(s) dg(s) \right\|$$
$$\leq \max\left(1, \frac{1}{\delta_1}\right) \int_0^T |\tilde{g}_1(t,s) - \tilde{g}_2(t,s)| \|f_1(s)\| dg(s)$$
$$+ \max\left(1, \frac{1}{\delta_1}\right) \max\left(1, \frac{1}{\delta_2}\right) \int_0^T (1 + m_2) |\tilde{g}_2(t,s)| \|f_1(s)\| |\alpha_1(s) - \alpha_2(s)| dg(s)$$
$$+ \max\left(1, \frac{1}{\delta_1}\right) \max\left(1, \frac{1}{\delta_2}\right) g(T) \int_0^T |\tilde{g}_2(t,s)| |b_1(s) - b_2(s)| \|f_1(s)\| dg(s).$$

We can also see, by the choice of f_2 that

$$\left\| \int_0^T \frac{\alpha_2(s)}{1 - b_2(s)\mu_g(\{s\})} \tilde{g}_2(t,s) (f_1(s) - f_2(s)) dg(s) \right\|$$
$$\leq \max\left(1, \frac{1}{\delta_2}\right) \int_0^T |\tilde{g}_2(t,s)| \cdot D(F_1(s), F_2(s)) dg(s).$$

We are now evaluating the second term of the sum (7):

$$\left\| (\alpha_1(t) - \alpha_2(t)) \int_0^T \frac{\alpha_2(s)}{1 - b_2(s)\mu_g(\{s\})} \tilde{g}_2(t,s) f_2(s) dg(s) \right\|$$

$$\leq \max\left(1, \frac{1}{\delta_2}\right) |\alpha_1(t) - \alpha_2(t)| \int_0^T \|\tilde{g}_2(t,s) f_2(s)\| dg(s).$$

As in Remark 4, we denote by

$$M_1 = \sup_{(s',s'') \in [0,T] \times [0,T]} e^{\int_{s'}^{s''} \tilde{b}_1(s) dg(s)},$$

respectively

$$M_2 = \sup_{(s',s'') \in [0,T] \times [0,T]} e^{\int_{s'}^{s''} \tilde{b}_2(s) dg(s)}$$

and so,

$$|\tilde{g}_1(t,s)| \leq \frac{\max(M_1, M_1^2)}{\left|\alpha_1(T) e^{\int_0^T \tilde{b}_1(r) dg(r)} - 1\right|} = \overline{M}_1$$

respectively

$$|\tilde{g}_2(t,s)| \leq \frac{\max(M_2, M_2^2)}{\left|\alpha_2(T) e^{\int_0^T \tilde{b}_2(r) dg(r)} - 1\right|} = \overline{M}_2.$$

It follows that

$$\|u_1(t) - u_2(t)\|$$
$$\leq \max\left(1, \frac{1}{\delta_1}\right) \int_0^T |\tilde{g}_1(t,s) - \tilde{g}_2(t,s)| \|f_1(s)\| dg(s)$$
$$+ \max\left(1, \frac{1}{\delta_1}\right) \max\left(1, \frac{1}{\delta_2}\right) \int_0^T (1 + m_2) |\tilde{g}_2(t,s)| \|f_1(s)\| |\alpha_1(s) - \alpha_2(s)| dg(s)$$
$$+ \max\left(1, \frac{1}{\delta_1}\right) \max\left(1, \frac{1}{\delta_2}\right) g(T) \int_0^T |\tilde{g}_2(t,s)| |b_1(s) - b_2(s)| \|f_1(s)\| dg(s)$$
$$+ \max\left(1, \frac{1}{\delta_2}\right) \int_0^T |\tilde{g}_2(t,s)| \cdot D(F_1(s), F_2(s)) dg(s)$$
$$+ \max\left(1, \frac{1}{\delta_2}\right) |\alpha_1(t) - \alpha_2(t)| \int_0^T \|\tilde{g}_2(t,s) f_2(s)\| dg(s)$$

$$\|u_1(t) - u_2(t)\| \leq \max\left(1, \frac{1}{\delta_1}\right) \int_0^T |\tilde{g}_1(t,s) - \tilde{g}_2(t,s)| \overline{\phi}(s) dg(s)$$
$$+ \max\left(1, \frac{1}{\delta_1}\right) \max\left(1, \frac{1}{\delta_2}\right) (1 + m_2) \overline{M}_2 \int_0^T \overline{\phi}(s) |\alpha_1(s) - \alpha_2(s)| dg(s)$$
$$+ \max\left(1, \frac{1}{\delta_1}\right) \max\left(1, \frac{1}{\delta_2}\right) \overline{M}_2 g(T) \int_0^T |b_1(s) - b_2(s)| \overline{\phi}(s) dg(s)$$
$$+ \max\left(1, \frac{1}{\delta_2}\right) \overline{M}_2 \int_0^T D(F_1(s), F_2(s)) dg(s)$$
$$+ \max\left(1, \frac{1}{\delta_2}\right) \overline{M}_2 |\alpha_1(t) - \alpha_2(t)| \int_0^T \overline{\phi}(s) dg(s). \qquad (8)$$

We are now evaluating the difference $\tilde{g}_1(t,s) - \tilde{g}_2(t,s)$. It can be seen that

$$\tilde{g}_1(t,s) - \tilde{g}_2(t,s)$$
$$= \begin{cases} \dfrac{\alpha_1(T)}{\alpha_1(T)e^{\int_0^T \tilde{b}_1(r)dg(r)} - 1} e^{\int_0^T \tilde{b}_1(r)dg(r) - \int_s^t \tilde{b}_1(r)dg(r)} - \dfrac{\alpha_2(T)}{\alpha_2(T)e^{\int_0^T \tilde{b}_2(r)dg(r)} - 1} e^{\int_0^T \tilde{b}_2(r)dg(r) - \int_s^t \tilde{b}_2(r)dg(r)}, \\ \qquad\qquad\qquad\qquad\qquad\qquad if\ 0 \leq s \leq t \leq T \\ \dfrac{1}{\alpha_1(T)e^{\int_0^T \tilde{b}_1(r)dg(r)} - 1} e^{-\int_s^t \tilde{b}_1(r)dg(r)} - \dfrac{1}{\alpha_2(T)e^{\int_0^T \tilde{b}_2(r)dg(r)} - 1} e^{-\int_s^t \tilde{b}_2(r)dg(r)}, \\ \qquad\qquad\qquad\qquad\qquad\qquad if\ 0 \leq t < s \leq T. \end{cases}$$

In the first case ($0 \leq s \leq t \leq T$),

$$|\tilde{g}_1(t,s) - \tilde{g}_2(t,s)|$$
$$\leq \frac{1}{|(\alpha_1(T)e^{\int_0^T \tilde{b}_1(r)dg(r)} - 1)(\alpha_2(T)e^{\int_0^T \tilde{b}_2(r)dg(r)} - 1)|}$$
$$\left[e^{\int_0^T \tilde{b}_1(r)dg(r)} e^{\int_0^T \tilde{b}_2(r)dg(r)} \left| e^{-\int_s^t \tilde{b}_1(r)dg(r)} - e^{-\int_s^t \tilde{b}_2(r)dg(r)} \right| \right.$$
$$\left. + \left| \alpha_1(T)e^{\int_0^T \tilde{b}_1(r)dg(r) - \int_s^t \tilde{b}_1(r)dg(r)} - \alpha_2(T)e^{\int_0^T \tilde{b}_2(r)dg(r) - \int_s^t \tilde{b}_2(r)dg(r)} \right| \right]$$
$$\leq \frac{1}{|(\alpha_1(T)e^{\int_0^T \tilde{b}_1(r)dg(r)} - 1)(\alpha_2(T)e^{\int_0^T \tilde{b}_2(r)dg(r)} - 1)|}$$
$$\left[e^{\int_0^T \tilde{b}_1(r)dg(r)} e^{\int_0^T \tilde{b}_2(r)dg(r)} \left| e^{-\int_s^t \tilde{b}_1(r)dg(r)} - e^{-\int_s^t \tilde{b}_2(r)dg(r)} \right| \right.$$
$$+ \left| \alpha_1(T) \left(e^{\int_0^T \tilde{b}_1(r)dg(r) - \int_s^t \tilde{b}_1(r)dg(r)} - e^{\int_0^T \tilde{b}_2(r)dg(r) - \int_s^t \tilde{b}_2(r)dg(r)} \right) \right|$$
$$\left. + |\alpha_1(T) - \alpha_2(T)| e^{\int_0^T \tilde{b}_2(r)dg(r) - \int_s^t \tilde{b}_2(r)dg(r)} \right]$$

and, by Remark 5(i),

$$|\tilde{g}_1(t,s) - \tilde{g}_2(t,s)|$$
$$\leq \frac{1}{|(\alpha_1(T)e^{\int_0^T \tilde{b}_1(r)dg(r)} - 1)(\alpha_2(T)e^{\int_0^T \tilde{b}_2(r)dg(r)} - 1)|}$$
$$\left[M_1 M_2 \max(M_1, M_2) \left| \int_s^t \tilde{b}_1(r)dg(r) - \int_s^t \tilde{b}_2(r)dg(r) \right| \right.$$
$$+ \max(M_1^2, M_2^2) \left| \int_0^T \tilde{b}_1(r)dg(r) - \int_s^t \tilde{b}_1(r)dg(r) - \int_0^T \tilde{b}_2(r)dg(r) + \int_s^t \tilde{b}_2(r)dg(r) \right|$$
$$\left. + |\alpha_1(T) - \alpha_2(T)| M_2^2 \right]$$
$$\leq \frac{1}{|(\alpha_1(T)e^{\int_0^T \tilde{b}_1(r)dg(r)} - 1)(\alpha_2(T)e^{\int_0^T \tilde{b}_2(r)dg(r)} - 1)|}$$
$$\left[\left(M_1 M_2 \max(M_1, M_2) + \max(M_1^2, M_2^2) \right) \int_0^T |\tilde{b}_1(s) - \tilde{b}_2(s)| dg(s) + |\alpha_1(T) - \alpha_2(T)| M_2^2 \right].$$

Similarly, in the second case ($0 \leq t < s \leq T$) it can be proved that

$$|\tilde{g}_1(t,s) - \tilde{g}_2(t,s)|$$
$$\leq \frac{1}{|(\alpha_1(T)e^{\int_0^T \tilde{b}_1(r)dg(r)} - 1)(\alpha_2(T)e^{\int_0^T \tilde{b}_2(r)dg(r)} - 1)|}$$
$$\left[(M_1 M_2 + \max(M_1, M_2)) \int_0^T |\tilde{b}_1(s) - \tilde{b}_2(s)| dg(s) + |\alpha_1(T) - \alpha_2(T)| M_1 M_2 \right].$$

Denoting by

$$\tilde{M}_1 = \max\left(M_1 M_2 \max(M_1, M_2) + \max(M_1^2, M_2^2), M_1 M_2 + \max(M_1, M_2)\right)$$

respectively

$$\tilde{M}_2 = \max\left(M_2^2, M_1 M_2\right),$$

we may say that for every $s, t \in [0, T]$,

$$|\tilde{g}_1(t,s) - \tilde{g}_2(t,s)| \leq \tilde{M}_1 \int_0^T |\tilde{b}_1(s) - \tilde{b}_2(s)| dg(s) + \tilde{M}_2 |\alpha_1(T) - \alpha_2(T)|. \tag{9}$$

We use next Remark 5(ii) and the fact that from (4), any $t \in D_g$ satisfies

$$|1 - b_1(t)\mu_g(\{t\})| > \delta_1 \quad \text{and} \quad |1 - b_2(t)\mu_g(\{t\})| > \delta_2$$

to see that for each $t \in D_g$,

$$\begin{aligned}
|\tilde{b}_1(t) - \tilde{b}_2(t)| &= \left|\frac{-\log|1 - b_1(t)\mu_g(\{t\})| + \log|1 - b_2(t)\mu_g(\{t\})|}{\mu_g(\{t\})}\right| \\
&\leq \max\left(\frac{1}{\delta_1}, \frac{1}{\delta_2}\right) |b_1(t) - b_2(t)|.
\end{aligned}$$

It is immediate that for each $t \in [0, T]$,

$$|\tilde{b}_1(t) - \tilde{b}_2(t)| \leq \max\left(1, \frac{1}{\delta_1}, \frac{1}{\delta_2}\right) |b_1(t) - b_2(t)|. \tag{10}$$

Finally, exploiting (8), (9), (10) we obtain that for all $t \in [0, T]$,

$$\begin{aligned}
&\|u_1(t) - u_2(t)\| \\
&\leq \max\left(1, \frac{1}{\delta_1}\right) \int_0^T \overline{\phi}(s) dg(s) \left(\tilde{M}_1 \int_0^T |\tilde{b}_1(s) - \tilde{b}_2(s)| dg(s) + \tilde{M}_2 |\alpha_1(T) - \alpha_2(T)|\right) \\
&+ \max\left(1, \frac{1}{\delta_1}\right) \max\left(1, \frac{1}{\delta_2}\right) (1 + m_2) \overline{M}_2 \int_0^T \overline{\phi}(s) |\alpha_1(s) - \alpha_2(s)| dg(s) \\
&+ \max\left(1, \frac{1}{\delta_1}\right) \max\left(1, \frac{1}{\delta_2}\right) \overline{M}_2 g(T) \int_0^T |b_1(s) - b_2(s)| \overline{\phi}(s) dg(s) \\
&+ \max\left(1, \frac{1}{\delta_2}\right) \overline{M}_2 \int_0^T D(F_1(s), F_2(s)) dg(s) \\
&+ \max\left(1, \frac{1}{\delta_2}\right) \overline{M}_2 |\alpha_1(t) - \alpha_2(t)| \int_0^T \overline{\phi}(s) dg(s)
\end{aligned}$$

so

$$\|u_1(t) - u_2(t)\|$$
$$\leq \tilde{M}_1 \int_0^T \overline{\phi}(s) dg(s) \cdot \max\left(1, \frac{1}{\delta_1}\right) \cdot \max\left(1, \frac{1}{\delta_1}, \frac{1}{\delta_2}\right) \cdot \int_0^T |b_1(s) - b_2(s)| dg(s)$$
$$+ \tilde{M}_2 \max\left(1, \frac{1}{\delta_1}\right) \int_0^T \overline{\phi}(s) dg(s) \cdot |\alpha_1(T) - \alpha_2(T)|$$
$$+ (1 + m_2)\overline{M}_2 \max\left(1, \frac{1}{\delta_1}\right) \max\left(1, \frac{1}{\delta_2}\right) \cdot \int_0^T \overline{\phi}(s) |\alpha_1(s) - \alpha_2(s)| dg(s)$$
$$+ \overline{M}_2 g(T) \max\left(1, \frac{1}{\delta_1}\right) \max\left(1, \frac{1}{\delta_2}\right) \cdot \int_0^T |b_1(s) - b_2(s)| \overline{\phi}(s) dg(s)$$
$$+ \overline{M}_2 \max\left(1, \frac{1}{\delta_2}\right) \cdot \int_0^T D(F_1(s), F_2(s)) dg(s)$$
$$+ \overline{M}_2 \max\left(1, \frac{1}{\delta_2}\right) \int_0^T \overline{\phi}(s) dg(s) \cdot |\alpha_1(t) - \alpha_2(t)|.$$

Denoting thus by

$$C_1 = \tilde{M}_1 \int_0^T \overline{\phi}(s) dg(s) \cdot \max\left(1, \frac{1}{\delta_1}\right) \cdot \max\left(1, \frac{1}{\delta_1}, \frac{1}{\delta_2}\right),$$
$$C_2 = \tilde{M}_2 \max\left(1, \frac{1}{\delta_1}\right) \int_0^T \overline{\phi}(s) dg(s), \quad C_3 = (1 + m_2)\overline{M}_2 \max\left(1, \frac{1}{\delta_1}\right) \max\left(1, \frac{1}{\delta_2}\right),$$
$$C_4 = \overline{M}_2 g(T) \max\left(1, \frac{1}{\delta_1}\right) \max\left(1, \frac{1}{\delta_2}\right),$$
$$C_5 = \overline{M}_2 \max\left(1, \frac{1}{\delta_2}\right), \quad C_6 = \overline{M}_2 \max\left(1, \frac{1}{\delta_2}\right) \int_0^T \overline{\phi}(s) dg(s)$$

one gets the required inequality. □

Consider now

$$\overline{C}_1 = \tilde{M}_1 \int_0^T \overline{\phi}(s) dg(s) \cdot \max\left(1, \frac{1}{\delta_1}, \frac{1}{\delta_2}\right)^2,$$
$$\overline{C}_4 = \max(\overline{M}_2, \overline{M}_1) g(T) \max\left(1, \frac{1}{\delta_1}\right) \max\left(1, \frac{1}{\delta_2}\right),$$
$$\overline{C}_5 = \max(\overline{M}_2, \overline{M}_1) \max\left(1, \frac{1}{\delta_1}, \frac{1}{\delta_2}\right).$$

Corollary 1. *Under the assumptions of Theorem 5, if for every $t \in [0, T]$*

$$1 - b_1(t) \mu_g(t) > 0 \quad \text{and} \quad 1 - b_2(t) \mu_g(t) > 0$$

then:

(i)
$$D_C(\mathcal{S}_1, \mathcal{S}_2) \leq \left(\overline{C}_1 g(T) + \overline{C}_4 \int_0^T \overline{\phi}(s) dg(s)\right) \cdot \|b_1 - b_2\|_C + \overline{C}_5 g(T) \cdot D_C(F_1, F_2).$$

(ii) if $\overline{\phi}$ is bounded,

$$D_{L^1}(\mathcal{S}_1, \mathcal{S}_2) \leq \left(\overline{C}_1 + \overline{C}_4 \sup_{t \in [0,T]} \overline{\phi}(t)\right) g(T) \|b_1 - b_2\|_{L^1} + \overline{C}_5 g(T) D_{L^1}(F_1, F_2).$$

Proof. Under the additional hypothesis on b_1 and b_2, it can be seen that $\alpha_1(t) = \alpha_2(t) = 1$ on the whole interval and so, Theorem 5 yields that for every $u_1 \in \mathcal{S}_1$ one can find $u_2 \in \mathcal{S}_2$ such that for all $t \in [0, T]$,

$$\|u_1(t) - u_2(t)\| \leq C_1 \int_0^T |b_1(s) - b_2(s)| dg(s) + C_4 \int_0^T |b_1(s) - b_2(s)| \overline{\phi}(s) dg(s)$$
$$+ C_5 \int_0^T D(F_1(s), F_2(s)) dg(s). \qquad (11)$$

(i) By taking the supremum in (11) over $t \in [0, T]$,

$$\|u_1 - u_2\|_C \leq \left(C_1 g(T) + C_4 \int_0^T \overline{\phi}(s) dg(s) \right) \cdot \|b_1 - b_2\|_C + C_5 g(T) \cdot D_C(F_1, F_2).$$

By the definition of the Pompeiu-excess, it follows that

$$e_C(\mathcal{S}_1, \mathcal{S}_2) \leq \left(C_1 g(T) + C_4 \int_0^T \overline{\phi}(s) dg(s) \right) \cdot \|b_1 - b_2\|_C + C_5 g(T) \cdot D_C(F_1, F_2)$$

and, by interchanging the roles of $\mathcal{S}_1$ and $\mathcal{S}_2$, one obtains the announced estimation.

(ii) If $\overline{\phi}$ is bounded, the inequality (11) implies that

$$\|u_1(t) - u_2(t)\| \leq \left(C_1 + C_4 \sup_{t \in [0,T]} \overline{\phi}(t) \right) \|b_1 - b_2\|_{L^1} + C_5 D_{L^1}(F_1, F_2)$$

By integrating it with respect to g on $[0, T]$ we get

$$\|u_1 - u_2\|_{L^1} \leq \left(C_1 + C_4 \sup_{t \in [0,T]} \overline{\phi}(t) \right) g(T) \|b_1 - b_2\|_{L^1} + C_5 g(T) D_{L^1}(F_1, F_2)$$

whence

$$e_{L^1}(\mathcal{S}_1, \mathcal{S}_2) \leq \left(C_1 + C_4 \sup_{t \in [0,T]} \overline{\phi}(t) \right) g(T) \|b_1 - b_2\|_{L^1} + C_5 g(T) D_{L^1}(F_1, F_2).$$

and the inequality comes from interchanging the roles of $\mathcal{S}_1$ and $\mathcal{S}_2$. □

Author Contributions: Both authors have equally contributed to this work. All authors have read and agreed to the published version of the manuscript.

Funding: This research received no external funding.

Acknowledgments: The authors are grateful to the three referees for their comments and suggestions.

Conflicts of Interest: The authors declare no conflict of interest.

References

1. Federson, M.; Mesquita, J.G.; Slavík, A. Measure functional differential equations and functional dynamic equations on time scales. *J. Diff. Equ.* **2012**, *252*, 3816–3847. [CrossRef]
2. Kurzweil, J. Generalized ordinary differential equations and continuous dependence on a parameter. *Czechoslov. Math. J.* **1957**, *7*, 418–449. [CrossRef]
3. Schwabik, Š. *Generalized Ordinary Differential Equations*; World Scientific: London, UK, 1992.
4. Satco, B.; Smyrlis, G. Periodic boundary value problems involving Stieltjes derivatives. *J. Fixed Point Theory Appl.* **2020**, *22*, 94. [CrossRef]
5. Pouso, R.L.; Rodriguez, A. A new unification of continuous, discrete, and impulsive calculus through Stieltjes derivatives. *Real Anal. Exch.* **2015**, *40*, 319–353. [CrossRef]

6. Monteiro, G.A.; Satco, B. Distributional, differential and integral problems: Equivalence and existence results. *Electron. J. Qual. Theory Differ. Equ.* **2017**, *7*, 1–26. [CrossRef]
7. Frigon, M.; Pouso, R.L. Theory and applications of first-order systems of Stieltjes differential equations. *Adv. Nonlinear Anal.* **2017**, *6*, 13–36. [CrossRef]
8. Pouso, R.L.; Márquez, I.A. General existence principles for Stieltjes differential equations with applications to mathematical biology. *J. Differ. Equations* **2018**, *264*, 5388–5407. [CrossRef]
9. Cichoń, M.; Satco, B. Measure differential inclusions—Between continuous and discrete. *Adv. Diff. Equ.* **2014**, *56*, 18. [CrossRef]
10. Di Piazza, L.; Marraffa, V.; Satco, B. Closure properties for integral problems driven by regulated functions via convergence results. *J. Math. Anal. Appl.* **2018**, *466*, 690–710. [CrossRef]
11. Di Piazza, L.; Marraffa, V.; Satco, B. Approximating the solutions of differential inclusions driven by measures. *Ann. Mat. Pura Appl.* **2019**, *198*, 2123–2140. [CrossRef]
12. Silva, G.N.; Vinter, R.B. Measure driven differential inclusions. *J. Math. Anal. Appl.* **1996**, *202*, 727–746. [CrossRef]
13. Candeloro, D.; Croitoru, A.; Gavrilut, A.; Sambucini, A.R. An Extension of the Birkhoff Integrability for Multifunctions. *Mediter. J. Math.* **2016**, *13*, 2551–2575. [CrossRef]
14. Castaing, C.; Godet-Thobie, C.; Truong, L.X.; Mostefai, F.Z. On a Fractional Differential Inclusion in Banach Space Under Weak Compactness Condition. In *Advances in Mathematical Economics Volume 20*; Springer: Singapore, 2016; pp. 23–75.
15. Precupanu, A.; Gavrilut, A. Set-valued Lebesgue and Riesz type theorems. *Ann. Alexandru Ioan Cuza Univ. Math.* **2013**, *59*, 113–128. [CrossRef]
16. Chen, J.; Tisdell, C.C.; Yuan, R. On the solvability of periodic boundary value problems with impulse. *J. Math. Anal. Appl.* **2007**, *331*, 902–912. [CrossRef]
17. Li, J.; Nieto, J.; Shen, J. Impulsive periodic boundary value problems of first-order differential equations. *J. Math. Anal. Appl.* **2007**, *325*, 226–236. [CrossRef]
18. Zhao, A.; Bai, Z. Existence of solutions to first-order impulsive periodic boundary value problems. *Nonlinear Anal. TMA* **2009**, *71*, 1970–1977. [CrossRef]
19. Boucherif, A.; Tisdell, C. Existence of periodic and non-periodic solutions to systems of boundary value problems for first-order differential with super-linear growth. *Appl. Math. Comput.* **2008**, *204*, 441–449. [CrossRef]
20. Li, G.; Xue, X. On the existence of periodic solutions for differential inclusions. *J. Math. Anal. Appl.* **2002**, *276*, 168–183. [CrossRef]
21. Dhage, B.C.; Boucherif, A.; Ntouyas, N.K. On periodic boundary value problems of first-order perturbed impulsive differential inclusions. *Electr. J. Diff. Equ.* **2004**, *2004*, 1–9.
22. Graef, J.R.; Henderson, J.; Ouahab, A. *Impulsive Differential Inclusions. A Fixed Point Approach*; de Gruyter: Berlin, Germany, 2013.
23. Cichoń, M.; Satco, B.; Sikorska-Nowak, A. Impulsive nonlocal differential equations through differential equations on time scales. *Appl. Math. Comp.* **2011**, *218*, 2449–2458. [CrossRef]
24. Atici, F.M.; Biles, D.C. First order dynamic inclusions on time scales. *J. Math. Anal. Appl.* **2004**, *292*, 222–237. [CrossRef]
25. Frigon, M.; Gilbert, H. Systems of first order inclusions on time scales. *Topol. Meth. Nonlin. Anal.* **2011**, *37*, 147–163.
26. Fraňková, D. Regulated functions. *Math. Bohem.* **1991**, *116*, 20–59. [CrossRef]
27. Saks, S. *Theory of the Integral. Monografie Matematyczne*; Mathematical Association: Warszawa, Poland, 1937.
28. Monteiro, G.A.; Slavik, A.; Tvrdy, M. Kurzweil-Stieltjes integral, Theory and its Applications. In *Series in Real Analysis*; World Scientific: Singapore, 2018; Volume 15.
29. Tvrdý, M. Differential and Integral Equations in the Space of Regulated Functions. Habil. Thesis, Mathematical Institute—Academy of Sciences of the Czech Republic, Praha, Czech Republic, 2001.
30. Gordon, R.A. The Integrals of Lebesgue, Denjoy, Perron and Henstock. In *Graduate Students in Mathematics*; American Mathematical Society: Providence, RI, USA, 1994; Volume 4.
31. Satco, B. Existence results for Urysohn integral inclusions involving the Henstock integral. *J. Math. Anal. Appl.* **2007**, *336*, 44–53. [CrossRef]

32. Satco, B. Nonlinear Volterra integral equations in Henstock integrability setting. *Electr. J. Diff. Equ.* **2008**, *39*, 1–9.
33. Young, W.H. On integrals and derivatives with respect to a function. *Proc. Lond. Math. Soc.* **1917**, *2*, 35–63. [CrossRef]
34. Aubin, J.P.; Frankowska, H. *Set-Valued Analysis*; Birkhäuser: Boston, MA, USA, 1990.
35. Castaing, C.; Valadier, M. Convex Analysis and Measurable Multifunctions. In *Lecture Notes in Mathematics*; Springer: Berlin/Heidelberg, Germany, 1977; Volume 580.
36. Hu, S.; Papageorgiou, N.S. *Handbook of Multivalued Analysis. Volume I: Theory*; Kluwer Academic Publishers: Dordrecht, The Netherlands, 1997.

Publisher's Note: MDPI stays neutral with regard to jurisdictional claims in published maps and institutional affiliations.

 © 2020 by the authors. Licensee MDPI, Basel, Switzerland. This article is an open access article distributed under the terms and conditions of the Creative Commons Attribution (CC BY) license (http://creativecommons.org/licenses/by/4.0/).

Article

On Regulated Solutions of Impulsive Differential Equations with Variable Times

Diana Caponetti [1], Mieczysław Cichoń [2] and Valeria Marraffa [1,*]

[1] Department of Mathematics and Computer Science, University of Palermo, Via Archirafi 34, 90123 Palermo, Italy; diana.caponetti@unipa.it
[2] Faculty of Mathematics and Computer Science, A. Mickiewicz University, ul. Uniwersytetu Poznańskiego 4, 61-614 Poznań, Poland; mcichon@amu.edu.pl
* Correspondence: valeria.marraffa@unipa.it

Received: 11 November 2020; Accepted: 1 December 2020; Published: 4 December 2020

Abstract: In this paper we investigate the unified theory for solutions of differential equations without impulses and with impulses, even at variable times, allowing the presence of beating phenomena, in the space of regulated functions. One of the aims of the paper is to give sufficient conditions to ensure that a regulated solution of an impulsive problem is globally defined.

Keywords: regulated function; solution set; discontinuous function; impulsive problem with variable times

MSC: 34K40; 34A37; 34K05; 34K45; 47H30

---

1. Introduction

In recent years, impulse theory has been significantly developed, especially in the cases of impulsive differential equations or differential inclusions with fixed moments; see the monographs of Lakshmikantham et al. [1], Samoilenko and Perestyuk [2] and Perestyuk et al. [3] and the references therein. The study of impulsive problems with variable times presents more difficulties due to the state-dependent impulses, and in a large part of the literature, a finite number of impulses are still allowed. Some extensions to impulsive differential equations with variable times have been done by Bajo and Liz [4] and Frigon and O'Regan [5,6], and in the multivalued case, for instance, by Baier and Donchev or Gabor and Grudzka [7–9]. In the case of impulses at variable times, a "beating phenomenon" may occur, i.e., a solution of the differential equation may hit a given barrier several times (including infinitely many times). Then we will be in the presence of "pulse accumulation" whenever a solution has an infinite number of pulses which accumulate to a finite time t^*. Impulsive differential equations or inclusions have applications in physics, engineering or biology where discontinuities, which can be seen as impulses, occur [3,10]. In this paper we consider a class of initial value problems (IVPs) for differential equations with impulses at variable times on $[a,b]$, allowing pulse accumulation:

$$\begin{cases} x'(t) = f(t, x(t)), & t \notin \tau(x) \\ x(a) = x_0, \\ x(t) - x(t^-) = I_l(x(t-)), & t \in \tau(x) \\ x(t^+) - x(t) = I_r(x(t)), & t \in \tau(x) \end{cases}$$

where $f : [a,b] \times \mathbb{R} \to \mathbb{R}$, if not otherwise stated, is a continuous function; $\tau(x) \subset [a,b]$ is at most countable; and $I_r, I_l : \mathbb{R} \to \mathbb{R}$ and $x_0 \in \mathbb{R}$. Our consideration is presented for single-valued problems, but it is still valid for multivalued problems, as can be observed in [3,11], eventually by using multivalued integration [12–14].

Note that for a given function x the set $\tau(x)$ need not be a singleton. We study the case of accumulation points for the set $\tau(x)$. For an interesting discussion in this topic; see [15], where necessary and sufficient conditions are given to assure pulse accumulation. For problems having more than one common point of a solution and a barrier sufficient conditions are described in [16] (Theorem 4) or [1–3,17].

In this paper we study impulsive IVPs in the space $G([a,b])$ of regulated functions, which seems to be the natural space of solutions for impulsive problems (see [18–20]), and we investigate properties of solutions as elements of this space. This allows us to cover and extend earlier approaches. Note that usual IVPs should be treated as impulsive problems with negligible jumps. In this case the space $C([a,b])$ or $C^1([a,b])$ are considered, and they are subspaces of $G([a,b])$. We should note that impulsive differential equations with varying times of impulses are treated in [21] (Section 5) as generalized ordinary differential equations, but accumulation points for the set of discontinuity points are not allowed and solutions are functions of bounded variation. In [22] BV solutions are expected for impulsive problems. This approach was initiated by Silva and Vinter for the study of optimality problems driven by impulsive controls, but this space is not a proper choice in our study, as we need to consider only operators preserving bounded variation of functions and the norm in $BV([a,b])$ is not directly related to the supremum norm in $C([a,b])$. One of our goals is to unify the study for impulsive and non-impulsive problems. In the literature, IVPs with impulses at finite and fixed times have been studied in the subspace $PC([a,b],t_1,t_2,...,t_k)$ of the space $PC([a,b])$ of piecewise continuous functions, so that the space of solutions depends on times of jumps. In [23,24] the case of finite number of jumps is considered and the space of solutions is independent on times of jumps. In case of impulsive problems with variable times of jumps (state dependent jumps), a new space $CJ_k([a,b])$ is considered in [8,9,11] (for multivalued problems); it is a good choice for problems having the property that every solution has exactly k jumps; still, the space of solution depends on the choice of impulsive problem. We generalize previous approaches; indeed we have (some inclusions are taken in the sense of isometric copies)

$$C^1([a,b]) \subset C([a,b]) \subset PC([a,b],t_1,...,t_k) \subset CJ_k([a,b]) \subset PC([a,b]) \subset G([a,b]).$$

One of the advantages is that we are able to cover the case of beating phenomenon, till now studied separately and in very particular cases.

The paper is organized as follows. In Section 2 we recall basic notions on the space $G([a,b])$, and introduce, as space of solutions, the subspace Z_{GL} of regulated functions which admit only left accumulation points and have a canonical decomposition. We consider impulsive IVPs and provide conditions on the barriers which guarantee that solutions are global. In particular, condition [B4] requires that the sum of jumps (left and right) is finite and this condition implies that any solution is continuable to the point b. In Section 3 we give the equivalent representation of impulsive IVPs by means of operators acting on the space of regulated functions, and in the remaining part of the section we provide sufficient conditions for [B4]. An example is given in Section 4. Finally, in Section 5 we compare our results with earlier ones.

2. Impulsive Problems, Regulated Functions and Barriers

We denote by $G([a,b])$ the space of all real-valued regulated functions x defined on the interval $[a,b]$; that is, $G([a,b])$ is the set of all $x : [a,b] \to \mathbb{R}$ such that there exist finite the right $x(t^+)$ and left $x(s^-)$ limits for every points $t \in [a,b)$ and $s \in (a,b]$. The space $G([a,b])$ is a Banach space when equipped with the supremum norm (see [25]). The space $C([a,b])$ of continuous functions and the space $BV([a,b])$ of functions of bounded variation on $[a,b]$ are proper subspaces of $G([a,b])$, so on $BV([a,b])$ the induced norm is considered. Every regulated function is bounded, has a countable set of discontinuities and is the limit of a uniformly convergent sequence of step functions (cf. [26]). Given a regulated function $x \in G([a,b])$ we denote its set of discontinuity points by $\tau(x)$; if necessary, we distinguish the points of left-discontinuity $\tau_L(x)$ and right-discontinuity $\tau_R(x)$.

The following result, being an immediate consequence of a result by Bajo [15] (Theorem 1), implies that we need to restrict ourselves to some subspaces of regulated functions. Some necessary properties of solutions are described in the lemma below. We focus our attention on the subspace of regulated functions, denoted by $G^L([a,b])$, of all $x \in G([a,b])$, for which $\tau(x)$ has at most a finite number of left accumulation points (see [B2] for a more precise formulation).

Lemma 1. *If $t^* \in [a,b]$ is an accumulation point for the set of discontinuity points $\tau(x)$ of a regulated function $x : [a,b] \to \mathbb{R}$, then the size of the jumps is convergent to 0 when $t_n \to t^*$; i.e.,*

$$\lim_{t \in \tau(x), t \to t^*-} |x(t) - x(t^*-)| = 0 \quad \text{and} \quad \lim_{t \in \tau(x), t \to t^*+} |x(t^*+) - x(t)| = 0.$$

Now for $x \in G^L([a,b])$, we denote the left and right jump functions, respectively, by

$$J_L(x)(t) = x(t) - x(t-) \quad \text{and} \quad J_R(x)(t) = x(t+) - x(t),$$

for $t \in [a,b]$, where $x(a^-) = x(a)$ and $x(b^+) = x(b)$. Moreover for $t \in [a,b]$ we define

$$H_L(x)(t) = \sum_{t_k \in \tau_L(x), a \leq t_k \leq t} J_L(x)(t_k)$$

and

$$H_R(x)(t) = \sum_{t_k \in \tau_R(x), a \leq t_k < t} J_R(x)(t_k)$$

with $H_R(x)(a) = 0$. In the case of a finite number of left accumulation points it is understood that we will calculate the sum of the series of jumps separately for each such a point. Thus, we allow for conditional convergence of series as well. The key point of the paper is to decompose such a class of regulated functions as a sum of continuous and steplike functions (cf. [27]). Denote by Z_{G^L} the subspace of $G^L([a,b])$ consisting of regulated functions for which the sums $H_L(x)(t)$ and $H_R(x)(t)$ are finite for each $t \in [a,b]$. Then a function $x \in Z_{G^L}$ can be uniquely written as the sum of a continuous function and a steplike function.

The functions $x_d, x_c : [a,b] \to \mathbb{R}$ defined by setting

$$x_d(t) = H_L(x)(t) + H_R(x)(t)$$

and

$$x_c(t) = x(t) - x_d(t)$$

for $t \in [a,b]$ are called discrete and continuous parts of x. We will refer to $x = x_d + x_c$ as to the canonical decomposition of x; such a decomposition is unique with $x_d(a) = x(a)$ (cf. also [28] (Theorem 3)). We observe that for $t \in \tau(x)$ we have $J_L(x)(t) = x_d(t) - x_d(t-)$ and $x_c(t) - x_c(t-) = 0$, and analogously $J_R(x)(t) = x_d(t+) - x_d(t)$ and $x_c(t+) - x_c(t) = 0$, and

$$-\infty < \sum_{t_k \in \tau_L(x), a \leq t_k \leq b} J_L(x)(t_k) + \sum_{t_k \in \tau_R(x), a \leq t_k < b} J_R(x)(t_k) < \infty.$$

Moreover all functions $x \in Z_{G^L}$ are characterized by the condition that $J_L(x), J_R(x) \in l_1([a,b])$.

The spaces $C([a,b])$ and $BV([a,b])$ both are subspaces of Z_{G^L}. Moreover, also the space $CJ_k([a,b])$ is a subspace of Z_{G^L}. Let us stress that the function x_d is of bounded variation, but x_c need not have this property. For the sake of completeness we have to recall that a decomposition is possible for any function $x \in G([a,b])$, but without uniqueness (see [27–29]).

Let us consider the IVP for differential equations with impulses at variable times on $[a,b]$

$$\begin{cases} x'(t) = f(t, x(t)), & t \notin \tau(x) \\ x(a) = x_0, \\ x(t) - x(t^-) = I_l(x(t-)), & t \in \tau(x) \\ x(t^+) - x(t) = I_r(x(t)), & t \in \tau(x), \end{cases} \quad (1)$$

where $f : [a,b] \times \mathbb{R} \to \mathbb{R}$, $\tau(x) \subset [a,b]$ is at most countable and $I_r, I_l : \mathbb{R} \to \mathbb{R}$, and $x_0 \in \mathbb{R}$. Here I_r and I_l describe right and left jumps when $x(t)$ "touch" the barrier τ; i.e., $t \in \tau(x)$. If we expect one-side continuous solutions (càdlàg functions, for instance), then I_l or I_r should be trivial.

As a barrier we will understand a curve of the plane $\tau = \{(t,x) : t = \alpha(s), x = \beta(s), s \in \mathbb{R}\}$ or simply the graph of an equation $x = \gamma(t)$ for $t \in [a,b]$. Therefore, $\tau(x) = \{t \in [a,b] : x(t-) \in \tau\}$, and the functions I_r and I_l describe, respectively, right and left jumps of a solution $x(t)$ in the point $t \in [a,b]$ for which $x(t-)$ "touches" the barrier τ.

Throughout, we will consider the following conditions:

[B1] The point $(a, x_0) \notin \tau$.

[B2] If the set $\tau(x)$, for a solution x of (1), is not finite, then $\tau(x)$ has at most a finite number of accumulation points. For any accumulation point t^* of $\tau(x)$ there is an increasing sequence $\{t_k\}_{k \in \mathbb{N}}$ in $\tau(x)$ such that $t_k \to t^*$ and $t \notin \tau(x)$ whenever $t \in (t_k, t_{k+1})$.

[B3] In case of presence of more than one barrier (or connected components of the barrier) τ_k, they should be disjoint sets on a plane ($\tau_k \cap \tau_j = \emptyset$ for $k \neq j$). These barriers will be always assumed to be piecewise continuous curves.

[B4] For any accumulation point t^* of $\tau(x)$ the jump functions I_r, I_l have locally bounded sums of jumps in t^*; i.e.,

$$-\infty < \sum_{t_k \in \tau(x), a \leq t_k < t^*} I_l(x)(t_k) + \sum_{t_k \in \tau(x), a \leq t_k < t^*} I_r(x)(t_k) < \infty. \quad (2)$$

Moreover, either τ is bounded or if a solution x has the property that $x(t_k) \to \infty$ for some $t_k \in \tau(x), k = 1, 2 \ldots$, then (2) holds with the sums taking over k.

Conditions [B1]–[B4] allow one to cover existing cases and to study the problem of the solvability of the impulsive differential equation in presence of the beating phenomenon. The first three assumptions are quite natural and are usually assumed in earlier papers. In particular, [B1] implies that we have always a time $t_1 > a$ such that $x(t-)$, for $t \in [a, t_1)$, does not touch the barrier. This enables us to propose a step-by-step procedure for $t_1 < t_2 < \ldots$ at least to the first accumulation point of $\tau(x)$. We observe that [B1] can be relaxed, if a is a point of discontinuity, then it should be isolated in $\tau(x)$ and we need to replace the initial condition $x(a) = x_0$ by $x(a+) = x_0$. In the sequel we are interested in obtaining sufficient conditions for [B4]; we point out that condition [B4] implies that any solution is continuable to the point b. In case of more than one barrier (or connected components of the barrier), it may happen than the jump functions can transfer points between them. Let us recall that we have two jump conditions and then when $x(t_1-) \in \tau_1$ we have the first left jump. Thus, if after the jump $x(t_1) \in \tau_2$, it is still not a reason to get again the new left jump (as $x(t_1-) \notin \tau_2$). Only the right jump occurs and $x(t_1+)$ is calculated as $x(t_1+) = x(t_1) + I_r(x(t_1))$. As we assume that a couple of actions for τ_1 is always required, it is the jump function associated with the first barrier τ_1, a trajectory continues with the new initial value condition $x(t_1+)$; i.e., the mapping does not jump twice or more than once at the same moment. Condition [B3] guarantees that any solution of (1) does not jump more than once at the same moment.

Definition 1. *A function $x \in G^L([a,b])$ is said to be a regulated solution of the impulsive IVP (1) if it is differentiable except at most countable set $\tau(x) = \{t_k : k \in \mathbb{N}\}$. Moreover, if $a \notin \tau(x)$, then x coincides with the interval $[a, t_1)$, where $t_1 = \min \tau(x)$, with the solution of the differential equation $z'(t) = f(t, z(t))$ with initial condition $z(0) = x_0$, and x coincides with the interval (t_k, t_{k+1}) with the solution of the differential* equation $z'(t) = f(t, z(t))$ with initial condition $z(t_k) = x(t_k+)$, and the function x satisfies, at the points of the set $\tau(x)$, jump conditions with functions I_l and I_r, respectively.

Remark 1. *If we expect only that $x \in AC((t_k, t_{k+1}))$ for $k \in \mathbb{N}$, i.e., differentiability a.e. on such intervals, then the above definition can be also considered (the Carathéodory case instead of continuous functions f). In the case of lack of jumps (i.e., for all x we get $\tau(x) = \emptyset$) we have C^1-solutions. In the case of the connected components of the barrier in the form of vertical lines $\tau(x) = \{t_1, t_2, ..., t_k\}$ for any x, we have piecewise continuous solutions. For the case of CJ_k-solutions we need to identify such solutions with regulated solutions with precisely k barriers, each of them describing exactly one point t_k, i.e., $\tau_k(x) = t_k$. Let us mention that even solutions being of bounded variation considered in some papers are also included in our class of regulated solutions.*

We look for regulated solutions globally defined on $[a, b]$. Let us consider the IVP of the ODE associated with (1)

$$\begin{cases} x'(t) = f(t, x(t)) \\ x(a) = x_0. \end{cases} \quad (3)$$

If for a given solution of the impulsive IVP (1) we have only a finite number of discontinuity points, then the solution is global iff the solution of the IVP (3) is so, and thus usual assumptions guaranteeing globality of solutions are sufficient for impulsive problems too. The case of countable number of discontinuity points for some solutions is more complicated. Indeed, as claimed in [3] (p. 9), it is not true that if a solution of the IVP (3) cannot be extended to some interval, then a solution of the impulsive IVP (1) cannot also be extended to the same interval. We will show that it depends rather on the barrier and jump functions than on the solution of the impulsive IVP. So it is important to have combined assumptions for the barrier and jump functions. Note that also the growth of the function f is important. Let us discuss the following example, modified from [17] (Example 3.1).

Example 1. *Consider the following (IVP) problem in $[0, 2\pi]$.*

$$\begin{cases} x'(t) = 1 & t \notin \tau(x) \\ x(0) = -\pi \\ x(t^+) - x(t) = I_r(x)(t) & t \in \tau(x), \\ x(t) - x(t^-) = I_l(x)(t) & t \in \tau(x), \end{cases}$$

where $\tau(x) = \arctan(x) + \pi$, $I_r(x)(t) \equiv 1$ and $I_l(x)(t) \equiv 0$. Clearly, the Cauchy problem $x'(t) = 1$, $x(0) = -\pi$ has unique solution $x(t) = t - \pi$ defined globally on $[0, 2\pi]$. This solution touches the barrier, for the first time for $t_1 = \pi$, so a jump occurs and we get $x(\pi+) = 1$ and the solution of the IVP is defined as $x(t) = t - \pi + 1$ up to the next point when its trajectory touch the barrier, say t_2. We can proceed with points t_k and we get $\lim_{k \to \infty} t_k = \frac{3\pi}{2}$, so we have an accumulation point for $\tau(x)$ and the solution of the IVP is not defined globally on $[0, 2\pi]$, despite the fact that Cauchy problem has a global solution.

Now consider the same problem with $I_r(x)(t) = \frac{1}{(x(t))^2+1}$. In this case we have the same solution on $[0, t_1]$ and even the first jump is the same and the next jumps are: $x(t_2+) - x(t_2) = \frac{1}{(x(t_2))^2+1}$, etc. We can also easily calculate the points t_k and we get $\sum_{k=1}^{\infty} [x(t_k+) - x(t_k)] = M < \infty$ and as $t_k \to T < \frac{3\pi}{2}$ and we can put $x(t) = x + M - T$ for $t \in [T, 2\pi]$. We still get a global solution for the impulsive problem.

3. Integral Form of Impulsive Problems

We will study impulsive problem (1), representing it by means of operators acting on the space of regulated functions. To this end, let us consider the operator F defined on the space $G^L([a, b])$ in the the following way:

$$F(x)(t) = x_0 + \int_a^t f(s, x(s))\, ds + \sum_{t_k \in \tau_L(x), a \le t_k \le t} I_l(x)(t_k-) + \sum_{t_k \in \tau_R(x), a \le t_k < t} I_r(x)(t_k). \quad (4)$$

Notice that for $x \in Z_{GL}, t \in [a,b]$, we have

$$\sum_{t_k \in \tau_L(x), a \leq t_k \leq t} I_l(x)(t_k-) = \sum_{a \leq s \leq t} I_l(x_d(s-)) \quad \text{and} \quad \sum_{t_k \in \tau_R(x), a \leq t_k < t} I_r(x)(t_k) = \sum_{a \leq s < t} I_r(x_d(s)).$$

The discrete part $F_d(x)$ of the operator F, which will depend only on x_d, has to preserve the finiteness of sums of jumps, whenever x_d has this property. This condition depends on the barrier and jump functions I_r, I_l. In case of pulse accumulation, their acting on barriers should decrease jumps and the corresponding conditions for jump functions should compensate possible divergence, so in the presence of pulse accumulation they should be rapidly decreasing in the neighborhood of such a point. We allow one to have a finite number of such points, and we will present some sufficient conditions guaranteeing that even in this case all solutions are global. In case of finite number of jumps there are no new restrictions. Let us observe that for any discontinuity point $t \in \tau(x)$ we have direct dependence of the values of both $x(t_k)$ and $x(t_k+)$ on the value $x(t_k-)$, so they also depend on the barrier τ considered in (1); indeed:

$$x(t_k+) = x(t_k) + I_r(x(t_k)) = x(t_k-) + I_l(x(t_k-)) + I_r[x(t_k-) + I_l(x(t_k-))]. \tag{5}$$

We will investigate operators on Z_{GL} of the following form:

$$F(x)(t) = x_0 + \int_a^t f(s, x(s))\, ds + \sum_{a \leq s \leq t} I_l(x_d(s-)) + \sum_{a \leq s < t} I_r(x_d(s)). \tag{6}$$

We need to check the existence of the integral, the convergence of discrete parts and that this decomposition is canonical. Some differentiability properties of x outside of $\tau(x)$ and finite limits on $\tau(x)$ are also necessary to be solutions of (1).

Proposition 1. *Assume that the conditions [B1]–[B3] hold true and that*

(F1) $f \in C([a,b] \times \mathbb{R})$;

(J1) *for any $x \in Z_{GL}$ and $t \in [a,b]$*

$$-\infty < \sum_{a \leq s \leq t} I_l(x_d(s-)) + \sum_{a \leq s < t} I_r(x_d(s)) < \infty.$$

Then F, defined in (6), maps Z_{GL} into itself. Moreover, the operator F has the unique canonical decomposition $F(x) = F_c(x) + F_d(x)$, with

$$F_c(x)(t) = x_0 + \int_a^t f(s, x(s))\, ds$$

and

$$F_d(x)(t) = \sum_{a \leq s \leq t} I_l(x_d(s-)) + \sum_{a \leq s < t} I_r(x_d(s)),$$

so $F_c(x)$ is the continuous part of $F(x)$ and $F_d(x)$ is its discrete part.

Proof. Let us recall that if $f \in C([a,b] \times \mathbb{R})$, the superposition operator $N_f(x)(t) = f(t, x(t))$ maps $G^L([a,b])$ into itself (cf. [30] (Theorem 3.1) and [31]). Hence, the operator F_c is well-defined and $F_c(x) \in C([a,b])$. Assumption (J1) implies that $F_d : Z_{GL} \to Z_{GL}$; since $F_c : Z_{GL} \to C([a,b])$, we have that F maps Z_{GL} into itself. Let $x \in Z_{GL}$ and decompose $F(x)$ canonically as $y_c + y_d$. We need to prove that $y_c = F_c(x)$ and $y_d = F_d(x)$. First we investigate the discrete part. As no jump occurs, due to [B1], at the point a we have $y_d(a) = 0 = F_d(x)(a)$. Clearly, both functions y_d and $F_d(x)$ should have exactly the same points of discontinuity. Thus, for $t \in [a, t_1)$ both are null functions. As $y(t_1-) = J_L(y)(t_1) = F_d(x)(t_1-)$ and $y(t_1+) = J_R(y)(t) = F_d(x)(t_1+)$ we get the same jumps at $t = t_1$, so the values $y(t_1)$ and $F_d(x)(t_1)$ are the same. Thus, the left limits at the next point of discontinuity, say t_2, are the same (both are equal to the right limits at t_1). Due to our assumption on the set of discontinuity points for x we can proceed until the endpoint of existence of both functions, so that $y_d = F_d(x)$. Then, $y_c = F(x) - y_d = F(x) - F_d(x) = F_c(x)$. □

It is important to provide a sufficient condition to check the assumption (J1) occurs (cf. also [B4]). Let us observe that we need to verify only the convergence of jumps at accumulation points t^* of sets $\tau(x)$. For an interesting discussion about the presence or absence of such points, see [15] or [32]. For a given solution function x, if the set $\tau(x)$ has no accumulation points and the barrier and jump functions are bounded, then it can be defined on a whole interval (global solutions) (cf. example in [15] (Remark)). If we allow it to have some accumulation points, the problem is much more complicated. We need to find some conditions ensuring that all solutions pass through the accumulation points of $\tau(x)$, so they are global and can be prolonged up to the point b (see [16,33], for instance). As the problem in a whole generality is very hard to be described, we restrict ourselves to one non-trivial jump function and to the barrier defined as the graph of a continuous function.

Example 2. *Let $f(t, x) = \frac{1}{\cos^2 t}$ for $0 \leq t < \frac{\pi}{2}$ and $f(t, x) = 0$ for $t \geq \frac{\pi}{2}$. Consider the following problem: $x'(t) = f(t, x)$, $x(0) = 0$, $I_l(u) = -1$, $I_r(u) = 0$ and $\gamma(t) \equiv 1$. It is easy to see that this problem has a unique solution x defined on $[0, \infty)$ with $\tau(x) = \arctan(\mathbb{N})$. Clearly, $\tau(x)$ has a left dense accumulation point $t = \frac{\pi}{2}$. Despite that γ and x are bounded and defined for all $t \geq 0$, the assumption [B4] is not satisfied and $x \in G^L([0, \frac{\pi}{2})) \setminus Z_{G^L}$ and $x \notin G^L([0, \frac{\pi}{2}])$.*

Let us present some extensions for [15] (Theorem 2) and (Corollary 1).

Proposition 2. *Let $f \in C([a, b] \times \mathbb{R})$, $\gamma : [a, b] \to \mathbb{R}$ be a continuous function, the barrier τ be the graph of $x = \gamma(t)$ and $I_l \in C(\mathbb{R}, \mathbb{R})$ be associated with γ. Let $t^* \in (a, b]$ and let x be a regulated solution of the problem (1) such that the point t^* is a left accumulation point for the set $\tau(x)$. Assume that the following conditions hold:*

1. *There exists a positive constant M such that $|f(t, x)| \leq M$ for all $t \in [a, b]$ and $x \in Z_{G^L}$;*
2. *The barrier τ satisfies [B1]–[B3];*
3. *γ is nonincreasing on the interval $(t^* - c, t^*)$ for some $c > 0$;*
4. *I_l is nondecreasing and $I_l(u) < 0$ for $u \in (\gamma(t_1), \gamma(t^*))$ and some $t_1 \in (t^* - c, t^*)$.*

Then $\sum_{a \leq s \leq t^*} -I_l(x_d(s-)) < \infty$ and x can be extended to the right of t^*, [B4] holds true, and so any solution of the problem belongs to Z_{G^L}.

Proof. Let x be a regulated solution of the impulsive problem (1) for which t^* is a left accumulation point of $\tau(x)$. Set $u^* = \gamma(t^*)$; then, due to the continuity of γ, the point $(t^*, u^*) \in \tau$. Let (t_k) be a sequence in $[a, t^*)$ convergent to t^*. Without loss of generality, we may assume that $t_1 > t^* - c$, so γ is nonincreasing on (l_1, l^*). Fix an arbitrary regulated solution x of the impulsive problem (1). Fix $k \in \mathbb{N}$. Denote $u_k = x(t_k-) - \gamma(t_k)$. Then $(t_k, u_k) \in \tau$. We can estimate the position of the next point. Consider the system of equations: $x = \gamma(t)$ and $x = M \cdot t + u_k + I_l(u_k) - M \cdot t_k$ and denote by t^*_{k+1} the first solution to right of t_k. Moreover, as $|x'(t)| = |f(t, x(t))| \leq M$ for $t \notin \tau(x)$, we also have $t^*_{k+1} \leq t_{k+1}$. Since t^*_{k+1} is a solution of the equation $M \cdot t + u_k + I_l(u_k) - M \cdot t_k = \gamma(t)$, using the fact that γ is nonincreasing, we have

$$M \cdot t_{k+1} + u_k - (-I_l(u_k)) - M \cdot t_k \geq M \cdot t^*_{k+1} + u_k - (-I_l(u_k)) - M \cdot t_k = \gamma(t^*_{k+1}) \geq \gamma(t_{k+1}) = u_{k+1}.$$

From the latter we deduce

$$M \cdot (t_{k+1} - t_k) - (u_{k+1} - u_k) \geq -I_l(u_k) > 0.$$

Thus, for any $N \geq 1$, we have

$$\sum_{k=1}^{N}(-I_l(u_k)) \leq M \cdot \sum_{k=1}^{N}(t_{k+1} - t_k) - \sum_{k=1}^{N}(u_{k+1} - u_k),$$

and passing to the limit, we obtain

$$\begin{aligned}
\sum_{k=1}^{\infty} -I_l(u_k) &= \sum_{k=1}^{\infty} -I_l(x(t_k-)) = \sum_{a \leq s \leq t} -I_l(x_d(s-)) \\
&\leq \lim_{N \to \infty} \left(M \cdot \sum_{k=1}^{N}(t_{k+1} - t_k) - \sum_{k=1}^{N}(u_{k+1} - u_k) \right) \\
&= \lim_{N \to \infty} M \cdot (t_{N+1} - t_1) - \lim_{N \to \infty}(u_{N+1} - u_1) = M \cdot (t^* - t_1) + (u_1 - u^*) < \infty.
\end{aligned}$$

□

The analogy of Proposition 2 holds when γ is nondecreasing.

Proposition 3. *Let $f \in C([a,b] \times \mathbb{R})$, $\gamma : [a,b] \to \mathbb{R}$ be a continuous function, the barrier τ be a the graph of $x = \gamma(t)$ and $I_l \in C(\mathbb{R}, \mathbb{R})$ be associated with γ. Let $t^* \in (a,b]$ and let x be a regulated solution of the problem (1) such that the point t^* is a left accumulation point for the set $\tau(x)$. Assume that conditions 1 and 2 of Proposition 2 hold true and also:*

3'. γ is nondecreasing on the interval $(t^* - c, t^*)$ for some $c > 0$;
4'. I_l is nonincreasing and $I_l(u) > 0$ for $u \in (\gamma(t_1), \gamma(t^*))$ and some $t_1 \in (t^* - c, t^*)$.

Then $\sum_{a \leq s \leq t^*} I_l(x_d(s-)) < \infty$ and x can be extended to the right of t^*, [B4] holds true and so any solution of the problem belongs to Z_{GL}.

Proof. In this case we have an equation $x = -M \cdot t + u_k + I_l(u_k) + M \cdot t_k$, and if t^*_{k+1} denotes a solution of the equation $-M \cdot t + u_k + I_l(u_k) + M \cdot t_k = \gamma(t)$, then

$$M \cdot (t_{k+1} - t_k) + (u_{k+1} - u_k) \geq I_l(u_k).$$

Arguing as above we obtain

$$\sum_{k=1}^{\infty} I_l(x(t_k-)) = \sum_{a \leq s \leq t} I_l(x_d(s-)) = \sum_{k=1}^{\infty} I_l(u_k) < \infty.$$

□

In view of (5) we can formulate similar sufficient conditions considering both left and right jump functions.

Theorem 3.1. *Let $f \in C([a,b] \times \mathbb{R})$, $\gamma : [a,b] \to \mathbb{R}$ be a continuous function, the barrier τ be the graph of $x = \gamma(t)$ and $I_l, I_r \in C(\mathbb{R}, \mathbb{R})$. Let $t^* \in (a,b]$ and let x be a regulated solution of the problem (1) such that the point t^* is a left accumulation point for the set $\tau(x)$. Assume that the following conditions hold:*

1. There exists a positive constant M such that $|f(t,x)| \leq M$ for all $t \in [a,b]$ and $x \in Z_{GL,r}$;
2. The barrier τ satisfies [B1]–[B3];
3. γ is nonincreasing on the interval $(t^* - c, t^*)$ for some $c > 0$;
4. I_l and I_r are nondecreasing and $I_l(u) < 0$, $I_r(u) < 0$ for $u \in (\gamma(t_1), \gamma(t^*))$ and some $t_1 \in (t^* - c, t^*)$.

Then (J1) holds true; i.e., $-\infty < \sum_{a \leq s \leq t} I_l(x_d(s-)) + \sum_{a \leq s < t} I_r(x_d(s)) < \infty$, and x can be extended to the right of t^*.

Proof. We consider the affine function:

$$x = M \cdot t + u_k + I_l(u_k) + I_r(u_k + I_l(u_k)) - M \cdot t_k$$

and we get similar estimation as in Proposition 2,

$$u_{k+1} = \gamma(t_{k+1}) \leq \gamma(t^*_{k+1}) = M \cdot t^*_{k+1} + u_k + I_l(u_k) + I_r(u_k + I_l(u_k)) - M \cdot t_k.$$

As $I_l(u_k) < 0$, then $u_k + I_l(u_k) < u_k$. Thus

$$-I_l(x(t_k-)) + I_r(x(t_k)) \leq M(t_{k+1} - t_k) + (u_k - u_{k+1}).$$

The convergence of the series can be deduced as previously. □

Remark 2. *An analogous result of the previous Theorem can be obtained considering hypotheses (3') and (4') of Proposition 3.*

Corollary 1. *Under the assumptions of Proposition 3.1 there exists constant A such that all solutions x of the IVP (1) have equi-bounded sums of jumps:*

$$\sum_{a \leq s \leq t} |I_l(x_d(s-))| + \sum_{a \leq s < t} |I_r(x_d(s))| \leq A.$$

Proof. We restrict ourselves to proving the result in the case of left jumps. Put $a_k = \gamma(t^*_k)$, where (t^*_k) is the sequence constructed in Proposition 2, and let $A = \sum_{k=1}^{\infty} a_k$. Observe that for any solution x points of jumps $t_k \geq t^*_k$, so by the property of γ we get $\gamma(t^*_k) \geq \gamma(t_k)$ and then $I_l(\gamma(t^*_k)) \geq I_l(\gamma(t_k))$. For any x we get $\sum_{a \leq s \leq t} I_l(x_d(s-)) = \sum_{k=1}^{\infty} I_l(u_k) \leq \sum_{k=1}^{\infty} a_k = A < \infty$. □

Finally, we show that existence of solutions of IVP (1) is equivalent to existence of fixed points of operator F defined in (6) that are solutions of the following integral equation:

$$x(t) = x_0 + \int_a^t f(s, x(s)) \, ds + \sum_{a \leq s \leq t} I_l(x(s-)) + \sum_{a \leq s < t} I_r(x(s)). \tag{7}$$

Theorem 3.2. *Assume that the conditions [B1]–[B3] hold true and conditions (F1) and (J1) are satisfied. Then a function $x : [a, b] \to \mathbb{R}$ is a regulated solution of problem (1) on $[a, b]$ if and only if it is a fixed point of the operator F given by (6), i.e., a regulated solution of the integral Equation (7).*

Proof. ($\Leftarrow$) Let x be a solution of (7). Due to Proposition 1 we know that it belongs to $Z_{GL} \subset Z_G \subset G([a, b])$ and has a decomposition into a continuous part $x_0 + \int_a^t f(s, x(s)) \, ds$ and a discrete part $\sum_{a \leq s \leq t} I_l(x(s-)) + \sum_{a \leq s < t} I_r(x(s))$.

Immediately, we get that x satisfies the initial condition. Let $t \in [a, b]$ be a point of continuity, i.e., $t \notin \tau(x)$. Then $x'(t) = (\int_a^t f(s, x(s)) \, ds)' = f(t, x(t))$ so the differential equation is satisfied at such a point t. Now, let $t \in \tau(x)$. Let us calculate the jumps at this point. We have

$$\begin{aligned} x(t) - x(t-) &= x_0 + \int_a^t f(s, x(s)) \, ds + \sum_{a \leq s \leq t} I_l(x_d(s-)) + \sum_{a \leq s < t} I_r(x_d(s)) \\ &\quad - \left[x_0 + \int_a^t f(s, x(s)) \, ds + \sum_{a \leq s < t} I_l(x_d(s-)) + \sum_{a \leq s < t} I_r(x_d(s)) \right] \\ &= I_l(x(t-)), \end{aligned}$$

so the jump is precisely described by the function I_l. For the right jump we have similar calculation, so that $x(t+) - x(t) = I_r(x(t))$.

($\Rightarrow$) Let x be a regulated solution of the problem (1). As the superposition $f(\cdot, x(\cdot))$ is again regulated (cf. Proposition 1), it is an integrable function. Then if $t \in [a,b]$ is a point of continuity, we get $(\int_a^t f(s, x(s))\, ds)' = x'(t)$.

Since its left and right jumps at the points $t \in \tau(x)$ are described by jump functions $I_l(x(t))$ and $I_r(x(t))$, respectively, then by the definition of the discrete part, x_d is a sum of jumps, so $x_d(t) = \sum_{a \leq s \leq t} I_l(x_d(s-)) + \sum_{a \leq s < t} I_r(x_d(s))$ and finally $x(t) = x_c(t) + x_d(t) = F_c(x)(t) + F_d(x)(t)$. □

Now, let us present some consequences of our approach to the theory of differential inclusions. We will restrict our attention to the case of impulsive differential inclusions considered, for example, in [34] or [10] (cf. also [8,9]):

$$\begin{cases} x'(t) \in F(t, x(t)), & t \notin \tau(x) \\ x(0) = x_0, \\ x(t) - x(t^-) = I_l(x(t)), & t \in \tau(x) \\ x(t^+) = x(t), & t \in \tau(x) \end{cases} \quad (8)$$

where $F: [0,1] \times \mathbb{R}^d \to \mathcal{P}_{ck}(\mathbb{R}^d)$ is a multifunction with compact non-necessarily convex values in a real Euclidean space. In order to draw the readers' attention especially to new aspects of the paper, and not to focus their attention on the concepts of multi-valued analysis, let us refer them to [34] for definitions from multivalued analysis which will be used here after. In our evidence, we will only focus on the application of the previously obtained results, and the remaining details can be found in the literature.

We need to recall that in [34] the jump condition is of the form

$$\Delta x|_{t=\tau_i(x)} = S_i(x), \quad i = 1, \ldots, p, \; x(t) \in \mathbb{R}^d. \quad (9)$$

By an R-solution we mean an absolutely continuous function on each (τ_i, τ_{i+1}) for $i = 0, 1, \ldots, p, p+1$ ($\tau_0 = 0$ and $\tau_{p+1} = 1$) with impulses $\Delta x|_{t=\tau_i(x)} = S_i(x(\tau_i(x)^-))$; i.e., $x(\tau_i(x)^+) = x(\tau_i(x)^-) + S_i(x(\tau_i(x)^-))$, which satisfy $x'(t) \in F(t, x(t))$, $x(0) = x_0$ with $t \neq \tau_i(x)$ and (9).

The definition of R-solutions is more general than continuous or piecewise continuous solutions, but still it is more restrictive than ours. Consequently, we are ready to prove some results under less restrictive assumptions. Indeed, from our point of view, the most restrictive assumptions are those relating to barriers (cf. [34] (Assumptions (A1) and (A2))), which implies existence of at most p points of discontinuity for any solution x. Clearly, any R-solution is a regulated one, but not conversely.

Let us present two immediate generalizations of Proposition 2.

Proposition 4. (cf. [34] (Theorem 2.3)) *Let $F: [0,1] \times \mathbb{R}^d \to \mathbb{R}^d$ be almost usc multifunction with convex (and compact) values. Assume that the following conditions hold:*

1. *There exists a constant C such that $|F(t,x)| \leq C$ for every x and a.e. $t \in [0,1]$.;*
2. *The barrier τ satisfies [B1]–[B3];*
3. *γ is nonincreasing on the interval $(t^* - c, t^*)$ for some $c > 0$, provided that the point t^* is a left accumulation point for the set $\tau(x)$ and for any continuous function x satisfying $x'(t) \in F(t, x(t))$ and $x(0) = x_0$;*
4. *I_l is nondecreasing and $I_l(u) < 0$ for $u \in (\gamma(t_1), \gamma(t^*))$ and some $t_1 \in (t^* - c, t^*)$.*

Then there exists at least one regulated solution x for (8) and all solutions for this problem are global, i.e., they can be extended up to the right endpoint of the interval.

Proof. The proof is quite classical, so we want to draw attention to the differences resulting from our approach and related to the new definition of regulated solutions. The boundedness of F (hypothesis (A3) of [34] (Theorem 2.3)) allows us to conclude that if $G_\varepsilon(t, x) = \overline{co}\, F(([t - \varepsilon, t + \varepsilon] \cap$

$[0,1]) \setminus A, x + \varepsilon \mathbb{B})$ then $|G_\varepsilon(t,x)| \leq C$, where A is a null set and $\mathbb{B} \subset \mathbb{R}^d$ is the open unit ball. Then the set of functions being solutions of the initial value problem $x' \in F(t,x)$, $x(0) = x_0$ is nonempty.

Let 0 be a point of impulse. Then we consider (8) with an initial condition $x_0 + I_r x(0)$. Consequently, one can suppose without loss of generality that 0 is not a point of discontinuity. Thus, the differential inclusion without impulses

$$\begin{cases} x'(t) \in F(t,x(t)) \ t \in [0,1] \text{ a.e.,} \\ x(0) = x_0 \end{cases}$$

has continuous solutions (and the set of such solutions is compact in in $C([0,1], \mathbb{R}^d)$. For any such function x, either its graph touches the barrier γ on a set $\tau(x)$ consisting of finite numer of points, so by classical procedure (cf. [2,3]) it can be prolonged up to the point 1, or there exists some left accumulation point t^* for the set of $\tau(x)$.

Now, we take a solution of the above problem on $[0, t_1]$, where $t_1 = \min \tau(x)$ (see Definition 1), and step by step we construct our regulated solution on the whole interval $[0, t^*]$. We can repeat our procedure presented in Section 3; i.e., by Proposition 2 we get a function from Z_{GL} defined to the right of the point t^*. Recall, that this procedure is one of the main goals of this paper.

This procedure replaces the original one from [34] without any additional assumptions guarantying solutions with a number of discontinuity points prescribed by additional assumptions. Moreover, Proposition 2 implies that any solution exists on the interval $[0, 1]$. □

Let us consider also the lower semicontinuous case. The main idea of how to change the proof is essentially the same as in previous proposition.

Proposition 5 (cf. [34] (Theorem 2.8)). *Let $F : [0,1] \times \mathbb{R}^d \to \mathbb{R}^d$ be an almost lower semi-continuous on $\mathcal{A}$, with some negligible set $\mathcal{A}$; $F(\cdot, x)$ is measurable for every x; $F(t, \cdot)$ is upper semi-continuous with convex values on $([0,1] \times \mathbb{R}^d) \setminus \mathcal{A}$.*

Assume that the following conditions hold:

1. There exists a constant C such that $|F(t,x)| \leq C$ for every x and a.e. $t \in [0,1]$;
2. The barrier τ satisfies [B1]–[B3];
3. γ is nonincreasing on the interval $(t^* - c, t^*)$ for some $c > 0$, provided that the point t^* is a left accumulation point for the set $\tau(x)$ and for any continuous function x satisfying $x'(t) \in F(t,x(t))$ and $x(0) = x_0$;
4. I_l is nondecreasing and $I_l(u) < 0$ for $u \in (\gamma(t_1), \gamma(t^*))$ and some $t_1 \in (t^* - c, t^*)$.

Then there exists at least one regulated solution x for (8) x and all solutions for this problem are global, i.e., they can be extended up to the right endpoint of the interval.

4. Example

We present an explanatory example. We consider a classical Cauchy problem without uniqueness with the impulsive "stopping condition" on the interval $[0, a]$. To show the idea, it is sufficient to consider only one surface $\tau(x)$ with the property, that any solution with its graph reaching this surface has a jump. Put $H(x)(t) = x(t) - J(x)$, where $J(x) = 0$ for $x \leq 1$ and $J(x) = 1$ for $x > 1$, so $\tau(x)$ is the set of points t with $x(t) - 1 = 0$. Clearly $H_c(x) = x$ and $H_d(x) = -J(x)$.

$$\begin{cases} x'(t) &= 2\sqrt{x(t)} & t \notin \tau(x) \\ x(0) &= x(0+) = 0 \\ x(t+) - x(t-) &= H_d(x(t)) & t \in \tau(x). \end{cases} \quad (10)$$

As claimed above, let us find all the positions and the number of the points of discontinuity, i.e., the set $\tau(x)$. This set is depending on a solution x and then earlier results are not applicable in such a case.

Let us consider the integral form of this problem with $F(x)(t) = \int_0^t 2\sqrt{x(s)}\,ds + H_d(x(t))$, with $x_0 = 0$. The operator F takes the set of regulated functions Z_G into itself. For any $x \in Z_G$ we know that $H_d(x_d)$ has uniformly bounded sums $\sum_{k=0}^N \sqrt{k}$, where N is the number of jumps for a solution x, i.e., provided this sum is still less than a.

I. First let us present a general form for an arbitrary solution of (10). Since we know the formulae for all the solutions for the Cauchy problem (without the impulse condition), i.e., a trivial one $x_0(t) \equiv 0$ and $x_C(t) = 0$ for $t \in [0, C] \subset [0, a]$ and $x_C(t) = (t - C)^2$ for $a \geq t > C$, we can easily describe the set S_0 of all solutions for (10). All the intervals are considered here as intersections with $[0, a]$; i.e., $t \leq a$. Clearly, if $x_0(t) \equiv 0$, then $x_0 \in S_0$. Consider now an arbitrary function x_C. For $t_1 = C + 1$ we have $x_C(t_1) = 1$, so, using our condition, the function is "stopped" and $x_C(t_1+) = 0$. In such a way, we are again in the axis $y = 0$ and we are able to continue our procedure. The solution could be zero till the next point C_{k+1} in which we take $x_C(t) = (t - C_{k+1})^2$ or up to a. That means, the solution need not be determined by selecting only one point C. Then, for any set $Q = \{C_k \in [0, a] : k \in K \subset \mathbb{N}\}$, satisfying $C_{k+1} \geq C_k + \sqrt{k}$ $(k \in K)$, we associate a function x_Q having the form $x_Q(t) = (t - C_k)^2$ with some intervals $(C_k, C_k + \sqrt{k}]$ for all $C_k \in Q$ and vanishing elsewhere. Since x_Q is a bounded and regulated function, $S_0 \subset Z_G \subset G([0, a], \mathbb{R})$.

II. Note that different solutions of the considered problem can have different number of discontinuity points. Clearly, we have also infinitely many continuous solutions of our problem ($x \equiv 0$ and all functions having values zero up to a point C_k for which $(t - C_k)^2 < 1$ for $t \in [C_k, a]$).

The strength of our approach is more visible when we consider multivalued problems. Such a case is of special interest for unifying continuous and discontinuous approaches. Consider a modified problem from the previous example with the differential inclusion

$$x'(t) \in \left\{ 0, 2\sqrt{x(t)} \right\}, \quad t \notin \tau(x),$$

with the same set of conditions for impulses. Now, for arbitrary solution of previously considered problem at any point of its trajectory we can either prolong it as a constant function or continue as in Example 4. However, all solutions, both continuous and discontinuous, are still in our space Z_G. The case of convexified values of the above multifunction can be studied in the same manner.

5. Remarks about an Earlier Approach

In [9] (cf. also [8]) the following multivalued impulsive problem was studied:

$$\begin{aligned} y'(t) &\in F(t, y(t)), \quad \text{for } t \in [0, a], t \neq \tau_j(y(t)), j = 1, \ldots, m, \\ y(0) &= y_0, \\ y(t^+) &= y(t) + I_j(y(t)), \quad \text{for } t = \tau_j(y(t)), j = 1, \ldots, m, \end{aligned} \quad (11)$$

where $F : [0, a] \times \mathbb{R}^N \to 2^{\mathbb{R}^N}$, $I_j : \mathbb{R}^N \to \mathbb{R}^N$, $j = 1, \ldots, m$, are given impulse functions, $\tau_j \in C^1(\mathbb{R}^N, \mathbb{R})$ with $0 < \tau_j(y) < a$, and $t_y = \{t | t = \tau_k(y(t))\}$. The hypersurface $t - \tau_j(y) = 0$ is called the j-th pulse hypersurface and will be denoted by σ_j. If for each $j = 1, \ldots, m$, τ_j is a different constant function, then impulses are in the fixed times.

The authors are looking for (discontinuous) solutions in a special space. Let $CJ_m([0,a]) := C([0,a]) \times (\mathbb{R} \times \mathbb{R}^N)^m$ with following interpretation: the element $(\varphi, (l_j, v_j)_{j=1}^m)$, where $l_j \in [0,a]$ we will interpret as the function with m jumps in the times j_k defined as follows:

$$\hat{\varphi}(t) := \begin{cases} \varphi(t), & 0 \leq t \leq l_{\sigma(1)}, \\ \varphi(t) + \sum_{i=1}^{j} v_{\sigma(i)}, & l_{\sigma(j)} < t \leq l_{\sigma(j+1)}, \\ \varphi(t) + \sum_{i=1}^{m} v_{\sigma(i)}, & l_{\sigma(m)} < t \leq a, \end{cases}$$

where σ is a permutation of $\{1, 2, \ldots, m\}$ such that $l_{\sigma(i)} \leq l_{\sigma(i+1)}$.

The authors announced a mutual correspondence between the functions on interval $[0,a]$ with m jumps and the sets $\{(\varphi, (l_j, v_j)_{j=1}^m) \in CJ_m([0,a]) : l_j < l_{j+1}\}$, with $\zeta \mapsto (\check{\zeta}, (l_j, I_j(\check{\zeta}(l_j)))_{j=1}^m)$, where the function $\check{\zeta}$ is ζ with reduced jumps, l_j is j-th time of jump and the function I_j is an impulse function.

The space $CJ_m([0,a])$ with the norm

$$\|(\varphi, (l_j, v_j)_{j=1}^m)\| := \sup_{t \in [0,a]} \|\varphi(t)\| + \sum_{j=1}^m (|l_j| + \|v_j\|)$$

is a Banach space. In our approach it means that the considered functions are sums of continuous parts and discrete parts having finite number of discontinuity points. As the nature of mutual correspondences is not investigated in [9], solutions of the considered problem are included in this space $CJ_m([0,a])$. Thus, the problem is defined on a subset of continuous functions and the solution set is in a different space. Our approach allows one to eliminate such a problem. In contrast to our approach, the number of discontinuity points for solutions is then prescribed.

It is worthwhile to stress that our approach is based on analytical rather than topological methods and can be easily used for differential problems of various types having discontinuous solutions.

Let us mention that the main result in [9] is devoted to investigate the structure of the set of solutions for (11), and it was proved that under some assumptions this set is an R_δ set in $CJ_m([0,a])$. Despite that it exceeds the scope of this paper, it is an interesting problem and will be studied. Let us mention one big difference: our approach allows one to study problems with numbers of jumps depending on the solutions, including possibly infinite numbers of jumps.

The key difference in both cases is that we do not expect that all solutions of the considered should have prescribed (finite) number of discontinuity points. In [8,9] the authors have a finite number of "barriers" such that any solution meets each barrier (exactly one time). This means that several technical assumptions on that curves are required (conditions (H1)–(H3) in [11], for instance). As claimed above (and in our Example 4), the solutions studied by us have neither finite numbers of discontinuity points, nor the same number and placements of these points. An added value is that the space of solutions is universal for all problems having discontinuous solutions.

As claimed in Section 3, the same idea of solutions for differential inclusions having limited number of (possible) discontinuity points indicated by barriers met at once can be found in [34] or [10]. The space of solutions considered there consists of all functions x which are L-Lipschitz on $[\tau_i(x)^+, \tau_{i+1}(x)]$ and have no more than p jump points $\tau_1(x) < \tau_2(x) < \cdots < \tau_p(x)$. Note that in general τ_i depends on x; i.e., the impulses are not fixed times. Clearly, all such solutions are regulated.

Remark 3. *We propose to treat all such problems in an unified manner. First, we need to choose a proper subspace of $G([a,b], Y)$ and to define an operator on this space. Then either we have already a decomposition of this operator in its continuous and discrete parts (defined as in the formulation of a problem), or we need to decompose it like in our main theorem.*

Author Contributions: Writing—original draft, D.C., M.C. and V.M. All authors have read and agreed to the published version of the manuscript.

Funding: This research has been partially supported by GNAMPA, protocol U-UFMBAZ-2018-000351.

Conflicts of Interest: The authors declare no conflict of interest.

References

1. Lakshmikantham, V.; Bainov, D.D.; Simeonov, P.S. *Theory of Impulsive Differential Equations*; World Scientific: Singapore, 1989; Volume 6.
2. Samoilenko, A.M.; Perestyuk, N.A. *Impulsive Differential Equations*; Vyshcha Shkola: Kiev, Ukraine, 1987. (In Russian)
3. Perestyuk, N.A.; Plotnikov, V.A.; Samoilenko, A.M.; Skripnik, N.V. *Differential Equations with Impulse Effects: Multivalued Right-Hand Sides with Discontinuities*; Walter de Gruyter: Berlin, Germany, 2011; Volume 40.
4. Bajo, I.; Liz, E. Periodic boundary value problem for first order differential equations with impulses at variable times. *J. Math. Anal. Appl.* **1996**, *204*, 65–73. [CrossRef]
5. Frigon, M.; O'Regan, D. Impulsive differential equations with variable times. *Nonlinear Anal.* **1996**, *26*, 1913–1922. [CrossRef]
6. Frigon, M.; O'Regan, D. First order impulsive initial and periodic problems with variable moments. *J. Math. Anal. Appl.* **1999**, *233*, 730–739. [CrossRef]
7. Baier, R.; Donchev, T. Discrete Approximation of Impulsive Differential Inclusions. *Numer. Funct. Anal. Optim.* **2010**, *4-6*, 653–678. [CrossRef]
8. Gabor, G. Differential inclusions with state-dependent impulses on the half-line: New Fréchet space of functions and structure of solution sets. *J. Math. Anal. Appl.* **2017**, *446*, 1427–1448. [CrossRef]
9. Gabor, G.; Grudzka, A. Structure of the solution set to impulsive functional differential inclusions on the half-line. *NoDEA Nonlinear Differ. Equ. Appl.* **2012**, *19*, 609–627. [CrossRef]
10. Plotnikov, V.A.; Kitanov, N.M. On continuous dependence of solutions of impulsive differential inclusions and impulsive control problems. *Cybern. Syst. Anal.* **2007**, *38*, 759–771. [CrossRef]
11. Grudzka, A.; Ruszkowski, S. Structure of the solution set to differential inclusions with impulses at variable times. *Electron. J. Differ. Equ.* **2015**, *114*, 1–16.
12. Candeloro, D.; Di Piazza, L.; Musiał, K.; Sambucini, A.R. Integration of multifunctions with closed convex values in arbitrary Banach spaces. *JCA* **2020**, *27*, 1233–1246.
13. Candeloro, D.; Di Piazza, L.; Musiał, K.; Sambucini, A.R. Multifunctions determined by integrable functions. *Int. J. Approx. Reason.* **2019**, *112*, 140–148. [CrossRef]
14. Cichoń, K.; Cichoń, M.; Satco, B. Differential inclusions and multivalued integrals. *Discuss. Math. Differ. Incl. Control Optim.* **2013**, *33*, 171–191. [CrossRef]
15. Bajo, I. Pulse accumulation in impulsive differential equations with variable times. *J. Math. Anal. Appl.* **1997**, *216*, 211–217. [CrossRef]
16. Hu, S.; Lakshmikantham, V.; Leela, S. Impulsive differential systems and the pulse phenomena. *J. Math. Anal. Appl.* **1989**, *137*, 605–612. [CrossRef]
17. Dishliev, A.; Dishlieva, K.; Nenov, S. *Specific Asymptotic Properties of the Solutions of Impulsive Differential Equations. Methods and Applications*; Academic Publication: Dordrecht, The Netherlands, 2012.
18. Di Piazza, L.; Marraffa, V.; Satco, B. Closure properties for integral problems driven by regulated functions via convergence results. *J. Math. Anal. Appl.* **2018**, *466*, 690–710. [CrossRef]
19. Di Piazza, L.; Marraffa, V.; Satco, B. Measure differential inclusions: Existence results and minimum problems. *Set Valued Var. Anal.* 2020. [CrossRef]
20. Tvrdý, M. Differential and Integral Equations in the Space of Regulated Functions, Habil. Habilitation Thesis, Academy of Sciences of the Czech Republic: Praha, Czech Republic, 2001.
21. Afonso, S.M.; Bonotto, E.D.; Federson, M.; Schwabik, S. Discontinuous local semiflows for Kurzweil equations leading to LaSalle's invariance principle for differential systems with impulses at variable times. *J. Differ. Equ.* **2011**, *250*, 2969–3001. [CrossRef]
22. Samsonyuk, O.N.; Timoshin, S.A. BV solutions of rate independent processes driven by impulsive controls. *IFAC-PapersOnLine* **2018**, *51*, 361–366. [CrossRef]

23. Ballinger, G.; Liu, X. Existence and continuability of solutions for differential equations with delays and state-dependent impulses. *Nonlinear Anal.* **2002**, *51*, 633–647.
24. Ballinger, G.; Liu, X. Continuous dependence on initial values for impulsive delay differential equations. *Appl. Math. Lett.* **2004**, *17*, 483–490.
25. Fraňková, D. Regulated functions. *Math. Bohem.* **1991**, *116*, 20–59. [CrossRef]
26. Drewnowski, L. On Banach spaces of regulated functions. *Comment. Math. Pr. Mat.* **2017**, *57*, 153–169. [CrossRef]
27. Goffman, C.; Moran, G.; Waterman, D. The structure of regulated functions. *Proc. Am. Math. Soc.* **1976**, *57*, 61–65. [CrossRef]
28. Moran, G. Continuous functions on countable compact ordered sets as sums of their increments. *Trans. Am. Math. Soc.* **1979**, *256*, 99–112. [CrossRef]
29. Moran, G. Of regulated and steplike functions. *Trans. Am. Math. Soc.* **1977**, *231*, 249–257. [CrossRef]
30. Michalak, A. On superposition operators in spaces of regular and of bounded variation functions. *Z. Anal. Anwend.* **2016**, *35*, 285–309. [CrossRef]
31. Cichoń, K.; Cichoń, M.; Metwali, M. On some parameters in the space of regulated functions and their applications. *Carpathian J. Math.* **2018**, *34*, 17–30.
32. Fu, X.; Li, X. New results on pulse phenomena for impulsive differential systems with variable moments. *Nonlinear Anal.* **2009**, *71*, 2976–2984. [CrossRef]
33. Dishlieva, K. Continuous dependence of the solutions of impulsive differential equations on the initial conditions and barrier curves. *Acta Math. Sci.* **2012**, *32*, 1035–1052. [CrossRef]
34. Dontchev, T. Impulsive differential inclusions with constraints. *Electr. J. Differ. Equ.* **2006**, *66*, 1–12.

Publisher's Note: MDPI stays neutral with regard to jurisdictional claims in published maps and institutional affiliations.

© 2020 by the authors. Licensee MDPI, Basel, Switzerland. This article is an open access article distributed under the terms and conditions of the Creative Commons Attribution (CC BY) license (http://creativecommons.org/licenses/by/4.0/).

Article

The Riemann-Lebesgue Integral of Interval-Valued Multifunctions

Danilo Costarelli [1], Anca Croitoru [2], Alina Gavriluţ [2] and Alina Iosif [3] and Anna Rita Sambucini [1,*]

[1] Department of Mathematics and Computer Sciences, University of Perugia, 1, Via Vanvitelli, 06123 Perugia, Italy; danilo.costarelli@unipg.it
[2] Faculty of Mathematics, University Alexandru Ioan Cuza, Bd. Carol I, No. 11, 700506 Iaşi, Romania; croitoru@uaic.ro (A.C.); gavrilut@uaic.ro (A.G.)
[3] Department of Computer Science, Information Technology, Mathematics and Physics, Petroleum-Gas University of Ploieşti, Bd. Bucureşti, No. 39, 100680 Ploieşti, Romania; emilia.iosif@upg-ploiesti.ro
* Correspondence: anna.sambucini@unipg.it

Received: 10 November 2020; Accepted: 16 December 2020; Published: 20 December 2020

Abstract: We study Riemann-Lebesgue integrability for interval-valued multifunctions relative to an interval-valued set multifunction. Some classic properties of the RL integral, such as monotonicity, order continuity, bounded variation, convergence are obtained. An application of interval-valued multifunctions to image processing is given for the purpose of illustration; an example is given in case of fractal image coding for image compression, and for edge detection algorithm. In these contexts, the image modelization as an interval valued multifunction is crucial since allows to take into account the presence of quantization errors (such as the so-called round-off error) in the discretization process of a real world analogue visual signal into a digital discrete one.

Keywords: Riemann-Lebesgue integral; interval valued (set) multifunction; non-additive set function; image processing

MSC: 28B20; 28C15; 49J53

1. Introduction

The theory of multifunctions is an important field of research. Since interval arithmetic, introduced by Moore in [1], it appears a natural option for handling the uncertainty in data and in sensor measurements, particular attention was addressed to the study of interval-valued multifunctions and multimeasures because of their applications in statistics, biology, theory of games, economics, social sciences and software, to keep track of rounding errors in calculations and of uncertainties in the knowledge of the exact values of physical and technical parameters (see for example [2–5]). In fact, since the uncertainty of information could affect an expert's opinion, the ability to consider the uncertainty information during the process could be very important, see for example [2–4,6–11] and the references therein.

However, in some recent papers, interval-valued multifunctions have been applied also to some new directions, involving signal and image processing. Digital images are in fact the result of a discretization of the reality; namely sampled version of a continuous signal. Hence, there are different sources of uncertainty and ambiguity to be considered when performing image processing tasks, see for example [12,13]. For instance, the applications of fractal image coding for image compression [14,15] is one of the topic in which interval-valued multifunctions have been applied. Clearly, image compression techniques [16] are very useful in order to speed up the processes of digital image transmission and to improve the efficiency of image storage for high dimensional databases [17]. Further, applications of interval-valued multifunctions to the implementation of edge detection algorithms can also be found (see e.g., [13,18]).

In the literature several methods of integration for functions and multifunctions have been studied extending the Riemann and Lebesgue integrals. In this framework a generalization of Riemann sums was given in [19–37] while another generalization is due to Kadets and Tseytlin [38], who introduced the absolute Riemann-Lebesgue $|RL|$ and unconditional Riemann-Lebesgue RL integrability, for Banach valued functions with respect to countably additive measures. They proved that in finite measure space, the Bochner integrability implies $|RL|$ integrability which is stronger than RL integrability that implies Pettis integrability. Regarding this last extension contributions are given also in [21,23,34,39].

In the last decade the study of non-additive set functions and multifunctions has recently received a wide recognition, (see also [3,9,10,40–46]). In this paper, motivated by the large number of fields in which the interval-valued multifunction can be applied, we introduce a new type of integral of an interval-valued multifunction G with respect to an interval-valued submeasure M with respect to the weak interval order relation introduced in [4] by Guo and Zhang. Although the construction procedure of the integral is similar to the one given in [34,38,39], the integral proposed is a generalization of it since we are concerned with the study of a Riemann-Lebesgue set-valued integrand with respect to an arbitrary interval-valued set function, not necessarily countably additive. So the novelty of this construction concerns not only the codomain of the integrands but also the non-additivity of the measure with respect to which they are integrated. The main results on this subject are Theorem 1, in which the additivity of the integral is proved even if the pair (G, M) does not satisfy this property; the monotonicity and the order continuity are established in Theorems 2 and 4 and a convergent result given in Theorem 5.

The paper is organized as follows: in Section 2 the basic concepts and terminology are introduced together with some remarks. In Section 3 we introduce the RL-integral of an interval-valued multifunction with respect to an interval valued subadditive multifunction and we provide a comprehensive treatment of the integration theory together with a comparison with other integrals defined in the same setting (Remark 8). An example of an application in image processing is given in Section 3.1. The applications concerning image processing discussed in the present paper is given for the purpose of illustration and is new. The main reason for which we discuss the above application is to provide examples and justifications of the uses of interval-valued multifunctions to concrete applications in Image Processing. The advantage of using the notion of interval-valued multifunction in signal analysis is that this formalism allows to include in a unique framework possible uncertainty or the noise on the evaluation of an image at any given pixel.

2. Preliminaries

Let S be a nonempty at least countable set, $\mathcal{P}(S)$ the family of all subsets of S and $\mathcal{A}$ a σ-algebra of subsets of S. The symbol $\mathbb{R}_0^+$ denotes, as usual, the set of non negative real numbers.

Definition 1 ([34], Definition 2.1).

(i) *A finite (countable) partition of S is a finite (countable) family of nonempty sets $P = \{A_i\}_{i=1,\ldots,n}$ ($\{A_n\}_{n\in\mathbb{N}}) \subset \mathcal{A}$ such that $A_i \cap A_j = \emptyset, i \neq j$ and $\bigcup_{i=1}^{n} A_i = S$ ($\bigcup_{n\in\mathbb{N}} A_n = S$).*

(ii) *If P and P' are two partitions of S, then P' is said to be finer than P, denoted by $P \leq P'$ (or $P' \geq P$), if every set of P' is included in some set of P.*

(iii) *The common refinement of two finite or countable partitions $P = \{A_i\}$ and $P' = \{B_j\}$ is the partition $P \wedge P' = \{A_i \cap B_j\}$.*

(iv) *A countable tagged partition of S if a family $\{(B_n, s_n), n \in \mathbb{N}\}$ such that $(B_n)_n$ is a partition of S and $s_n \in B_n$ for every $n \in \mathbb{N}$.*

We denote by $\mathcal{P}$ the class of all the countable partitions of S and if $A \in \mathcal{A}$ is fixed, by $\mathcal{P}_A$ we denote the class of all the countable partitions of the set A.

Definition 2 ([34], Definition 2.2). *Let $m : \mathcal{A} \to [0, +\infty)$ be a non-negative function, with $m(\emptyset) = 0$. A set $A \in \mathcal{A}$ is said to be an atom of m if $m(A) > 0$ and for every $B \in \mathcal{A}$, with $B \subset A$, it is $m(B) = 0$ or $m(A \setminus B) = 0$.*

m is said to be:

(i) *monotone if $m(A) \leq m(B)$, $\forall A, B \in \mathcal{A}$, with $A \subseteq B$;*
(ii) *subadditive if $m(A \cup B) \leq m(A) + m(B)$, for every $A, B \in \mathcal{A}$, with $A \cap B = \emptyset$;*
(iii) *a submeasure (in the sense of Drewnowski [47]) if m is monotone and subadditive;*
(iv) *σ-subadditive if $m(A) \leq \sum_{n=0}^{+\infty} m(A_n)$, for every sequence of (pairwise disjoint) sets $(A_n)_{n \in \mathbb{N}} \subset \mathcal{A}$, with $A = \bigcup_{n=0}^{+\infty} A_n$.*
(v) *order-continuous (shortly, o-continuous) if $\lim_{n \to \infty} m(A_n) = 0$, for every decreasing sequence of sets $(A_n)_{n \in \mathbb{N}} \subset \mathcal{A}$, with $A_n \searrow \emptyset$;*
(vi) *exhaustive if $\lim_{n \to \infty} m(A_n) = 0$, for every sequence of pairwise disjoint sets $(A_n)_{n \in \mathbb{N}} \subset \mathcal{A}$.*
(vii) *null-additive if $m(A \cup B) = m(A)$, for every $A, B \in \mathcal{A}$, with $m(B) = 0$;*

Moreover m satisfies property (σ) if the ideal of m-zero sets is stable under countable unions (see for example [34], Definition 2.3).

We denote by the symbol $ck(\mathbb{R})$ the family of all non-empty convex compact subsets of $\mathbb{R}$, by convention, $\{0\} = [0,0]$. We consider on $ck(\mathbb{R})$ the Minkowski addition ($A + B := \{a + b : a \in A, b \in B\}$) and the standard multiplication by scalars. $\|A\| := \sup\{|x| : x \in A\}$. d_H is the Hausdorff distance in $ck(\mathbb{R})$, while $e(A, B) = \sup\{d(x, B), x \in A\}$ and $d_H(A, B) = \max\{e(A, B), e(B, A)\}$.

$(ck(\mathbb{R}), d_H)$ is a complete metric space ([48,49]), but is not a linear space since the subtraction is not well defined.

If $A = [a, b]$ then $\|A\| = \max\{|a|, |b|\}$. Moreover

$$d_H([a,b],[c,d]) = \max\{|a-c|, |b-d|\}, \quad \forall a, b, c, d \in \mathbb{R}$$
$$d_H([0,a],[0,b]) = |b-a| \quad \forall a, b \in \mathbb{R}_0^+.$$

In the family $ck(\mathbb{R})$ the following operations are also considered, for every $a, b, c, d \in \mathbb{R}$:

(i) $[a, b] \cdot [c, d] = [ac, bd]$;
(ii) $[a, b] \subseteq [c, d]$ if and only if $c \leq a \leq b \leq d$;
(iii) $[a, b] \preceq [c, d]$ if and only if $a \leq c$ and $b \leq d$; (weak interval order)
(iv) $[a, b] \wedge [c, d] = [\min\{a, c\}, \min\{b, d\}]$;
(v) $[a, b] \vee [c, d] = [\max\{a, c\}, \max\{b, d\}]$.

In general there is no relation between "$\preceq$" (iii) and "$\subseteq$" (ii); they only coincide on the subfamily $\{[0, a], a \geq 0\}$. Let $ck(\mathbb{R}_0^+) := \{[a, b], a, b \in \mathbb{R} \text{ and } 0 \leq a \leq b\}$.

In this paper we consider $(ck(\mathbb{R}_0^+), d_H, \preceq)$, namely the space $ck(\mathbb{R}_0^+)$ is endowed with the Hausdorff distance and the weak interval order. As a particular case of [20] (Definition 2.1) we have:

Definition 3. *Let $(a_n)_n$, $(b_n)_n$ be two sequences of real numbers so that $0 \leq a_n \leq b_n, \forall n \in \mathbb{N}$.*

The series $\sum_{n=0}^{\infty}[a_n, b_n] := \{\sum_{n=0}^{\infty} y_n : a_n \leq y_n \leq b_n, \forall n \in \mathbb{N}\}$ is called convergent if the sequence of partial sums $S_n := [\sum_{k=0}^{n} a_k, \sum_{k=0}^{n} b_k]$ is d_H-convergent to it.

Remark 1. *It is easy to see that $\sum_{n=0}^{\infty}[a_n, b_n] = [u, v]$, with $0 \leq u \leq v < \infty$, if and only if $\sum_{n=0}^{\infty} a_n = u$ and $\sum_{n=0}^{\infty} b_n = v$.*

We recall the following definition for the integrable Banach-valued functions $f : S \to X$ with respect to non-negative measures given in [38,39]:

Definition 4. *A function f is called unconditional Riemann-Lebesgue (RL) m-integrable (on S) if there exists $b \in X$ such that for every $\varepsilon > 0$, there exists a countable partition P_ε of S, so that for every countable partition $P = \{A_n\}_{n \in \mathbb{N}}$ of S with $P \geq P_\varepsilon$, f is bounded on every A_n, with $m(A_n) > 0$ and for every $t_n \in A_n$, $n \in \mathbb{N}$, the series $\sum_{n=0}^{+\infty} f(t_n)m(A_n)$ is unconditional convergent and*

$$\| \sum_{n=0}^{+\infty} f(t_n)m(A_n) - b \| < \varepsilon.$$

The vector b (necessarily unique) is called *the Riemann-Lebesgue m-integral of f on S* and it is denoted by $(RL)\int_S f dm$. The RL definition of the integrability on a subset $A \in \mathcal{A}$ is given in the classical manner.

Remark 2. *We remember that, in the countably additive case, unconditional RL-integrability is stronger than Birkhoff integrability (in the sense of Fremlin), see Ref. [23] and the references therein; while the notion of unconditional Riemann-Lebesgue integrability coincides with Birkhoff's one given in [21] (Definition 1, Proposition 2.6 and note at p. 8).*

For the properties of this integral with respect to a submeasure we refer to the results given in [34]. Moreover we have that

Proposition 1. *Let $g_n : S \to \mathbb{R}_0^+$ be an increasing sequence of bounded RL integrable function with respect to a submeasure $\mu : \mathcal{A} \to \mathbb{R}_0^+$ of bounded variation. If there exists a $g : S \to \mathbb{R}_0^+$ such that*

(a) $g_n \to g$ uniformly,
(b) $\sup_n (RL)\int_S g_n d\mu < +\infty$,

then g is RL integrable with respect to μ and

$$\lim_{n \to \infty} (RL)\int_S g_n \, d\mu = (RL)\int_S g \, d\mu.$$

Proof. Since $g_n \uparrow$, by the monotonicity we have that $(RL)\int_S g_n d\mu \uparrow$ so $\sup_n (RL)\int_S g_n \, d\mu = \lim_{n \to \infty} (RL)\int_S g_n \, d\mu = u \in \mathbb{R}_0^+$. Thanks to uniform convergence g is bounded; let $L > 0$ an upper bound for g.

Let $\varepsilon > 0$ be fixed and consider $k(\varepsilon) \in \mathbb{N}$ be such that

$$|g(t) - g_{k(\varepsilon)}(t)| < \frac{\varepsilon}{3\mu(S)} \quad \forall \, t \in S, \text{ and}$$

$$\left| (RL)\int_S g_{k(\varepsilon)} d\mu - u \right| < \frac{\varepsilon}{3}.$$

For every countable partition $P := (A_n)_n$ finer than $P_{\varepsilon/3,k(\varepsilon)}$ (the one that verifies Definition 4 for $g_{k(\varepsilon)}$) and for every $t_n \in A_n$ we have that $\sum_{n=0}^{+\infty} g(t_n)\mu(A_n)$ converges, since μ is of bounded variation.

In fact $g(t_n)\mu(A_n) \leq L\mu(A_n)$ for every $n \in \mathbb{N}$ and, for every $k \in \mathbb{N}$, it is $0 \leq \sum_{n=0}^{k} \mu(A_n) \leq \overline{\mu}(S)$. Moreover

$$\left| \sum_{n=0}^{+\infty} g(t_n)\mu(A_n) - u \right| \leq \left| \sum_{n=0}^{+\infty} g(t_n)\mu(A_n) - \sum_{n=0}^{+\infty} g_{k(\varepsilon)}(t_n)\mu(A_n) \right| +$$

$$+ \left| \sum_{n=0}^{+\infty} g_{k(\varepsilon)}(t_n)\mu(A_n) - (RL)\int_S g_{k(\varepsilon)} d\mu \right| +$$

$$+ \left| (RL)\int_S g_{k(\varepsilon)} d\mu - u \right| \leq \varepsilon.$$

□

Remark 3. *We can extend Proposition 1 to the bounded sequences $(g_n)_n$ that converge μ-almost uniformly on S (namely to the sequences $(g_n)_n$ such that for every $\varepsilon > 0$ there exists $B(\varepsilon) \in \mathcal{A}$ with $\mu(B(\varepsilon)) \leq \varepsilon$ and g_n converges uniformly to g on $S \setminus B(\varepsilon)$), if we assume that even g is bounded.*

We can proceed in fact in the same way, as in the previous proof, taking $P_\varepsilon^* := P_{\varepsilon/3, k(\varepsilon)} \wedge \{S \setminus B(\varepsilon), B(\varepsilon)\}$ and, for every countable partition $P := (A_n)_n$ finer than P_ε^*, dividing $\sum_{n=0}^{+\infty} g(t_n)\mu(A_n)$ in two parts: the one relative to $S \setminus B(\varepsilon)$, where the uniform convergence is assumed, and the remining part.

Convergence results in Gould integrability of functions with respect to a submeasure of finite variation are established for instance in [50].

Given two submeasures $\mu_1, \mu_2 : \mathcal{A} \to \mathbb{R}_0^+$ with $\mu_1(A) \leq \mu_2(A)$ for every $A \in \mathcal{A}$ let $M : \mathcal{A} \to ck(\mathbb{R}_0^+)$ defined by

$$M(A) = [\mu_1(A), \mu_2(A)]. \tag{1}$$

M is called an *interval submeasure*. For results in this subject see for example [3,43].

Let $M : \mathcal{A} \to ck(\mathbb{R}_0^+)$. We say that M is an *interval valued multisubmeasure* if

- $M(\emptyset) = \{0\}$;
- $M(A) \preceq M(B)$ for every $A, B \in \mathcal{A}$ with $A \subseteq B$ (monotonicity);
- $M(A \cup B) \preceq M(A) + M(B)$ for every disjoint sets $A, B \in \mathcal{A}$ (subadditivity).

In literature the multimeasures that satisfy the first two statements are also called set valued fuzzy measures (see for example [4] (Definition 1), [3,11,42–44] and the references therein).

A very interesting case of interval-valued multisubmeasure was given, for the first time, in [6,8] where Dempster and Shefer proposed a mathematical theory of evidence using non additive measures: Belief and Plausibility in such a way for every set A the *Belief interval* of the set is $[Bel(A), Pl(A)]$. This theory is capable of deriving probabilities for a collection of hypotheses and it allows the system inferencing with the imprecision and uncertainty. If the target space is $ck([0,1])$ it is used for example in decision theory.

We say that M is an *additive multimeasure* if $M(A \cup B) = M(A) + M(B)$ for every disjoint sets $A, B \in \mathcal{A}$.

If a multimeasure M is countably additive in the Hausdorff metric d_H, then it is called a d_H-*multimeasure*. In this case we have that $\lim_{n \to \infty} d_H(\sum_{k=1}^n M(A_k), M(A)) = 0$, for every sequence of pairwice disjoint sets $(A_n)_n \subset \mathcal{A}$ such that $\cup_n A_n = A$.

Remark 4. *By Ref. [43] (Remark 3.6) $M(A) = [\mu_1(A), \mu_2(A)]$ is a multisubmeasure with respect to $\preceq$ if and only if μ_1, μ_2 are submeasures in the sense of Definition 2 (iii). Moreover M is monotone, finitely additive, order-continuous, exhaustive respectively if and only if the set functions μ_1 and μ_2 are the same (see [40] (Proposition 2.5, Remark 3.3)).*

Definition 5. Let $M : \mathcal{A} \to ck(\mathbb{R}_0^+)$. The variation of M is the set function $\overline{M} : \mathcal{P}(S) \to [0, +\infty]$ defined by

$$\overline{M}(E) = \sup\{\sum_{i=1}^n \|M(A_i)\|,\ \{A_i\}_{i=1}^n \subset \mathcal{A},\ A_i \subseteq E, A_i \cap A_j = \emptyset, i \neq j\}.$$

M is said to be of finite variation if $\overline{M}(S) < \infty$.

Remark 5. We can observe that if $E \in \mathcal{A}$, then in the definition of $\overline{M}$ one may consider the supremum over all finite partitions $\{A_i\}_{i=1}^n \in \mathcal{P}_E$. If M is finitely additive, then $\overline{M}(A) = M(A)$, for every $A \in \mathcal{A}$.

If M is subadditive (countably subadditive, respectively) of finite variation, then $\overline{M}$ is finitely additive (countably additive, respectively). Finally, if $M(A) = [\mu_1(A), \mu_2(A)]$, for every $A \in \mathcal{A}$, then $\overline{M} = \overline{\mu}_2$.

3. RL Interval Valued Integral and Its Properties

In this section, we introduce and study Riemnn-Lebesgue integrability of interval-valued multifunctions with respect to interval-valued set multifunctions, pointing out various properties of this integral. For this, unless stated otherwise, in what follows suppose S is a nonempty set, with card $S \geq \aleph_0$ (card S is the cardinality of S), $\mathcal{A}$ is a σ-algebra of subsets of S.

The multisubmeasure M here considered is an interval-valued one and satisfies (1).

Given $g_1, g_2 : S \to \mathbb{R}_0^+$ with $g_1(s) \leq g_2(s)$ for all $s \in S$, let $G : S \to ck(\mathbb{R}_0^+)$ be the interval-valued multifunction defined by $G(s) = [g_1(s), g_2(s)]$ for every $s \in S$. For every countable tagged partition $\Pi := \{(B_n, s_n), n \in \mathbb{N}\}$ of S we denote by

$$\sigma_{G,M}(\Pi) := \sum_{n=1}^\infty G(s_n) \cdot M(B_n) = \sum_{n=1}^\infty [g_1(s_n)\mu_1(B_n), g_2(s_n)\mu_2(B_n)] =$$
$$= \{\sum_{n=1}^\infty y_n,\ y_n \in [g_1(s_n)\mu_1(B_n), g_2(s_n)\mu_2(B_n)],\ n \in \mathbb{N}\}.$$

By [20] (Lemma 2.2) the set $\sigma_{G,M}(\Pi)$ is closed and convex in $\mathbb{R}_0^+$, so it is an interval $[u_{G,M}^{(\Pi)}, v_{G,M}^{(\Pi)}]$.

Definition 6. A multifunction $G : S \to ck(\mathbb{R}_0^+)$ is called Riemann-Lebesgue RL integrable with respect to M (on S) if there exists $[a, b] \in ck(\mathbb{R}_0^+)$ such that for every $\varepsilon > 0$, there exists a countable partition P_ε of S, so that for every tagged partition $P = \{(A_n, t_n)\}_{n \in \mathbb{N}}$ of S with $P \geq P_\varepsilon$, the series $\sigma_{G,M}(P)$ is convergent and

$$d_H(\sigma_{G,M}(P), [a, b]) < \varepsilon. \tag{2}$$

$[a, b]$ is called the Riemann-Lebesgue integral of G with respect to M and it is denoted

$$[a, b] = (RL)\int_S G\, dM.$$

Obviously, if it exists, is unique.

Example 1. Suppose $S = \{s_n | n \in \mathbb{N}\}$ is countable, $\{s_n\} \in \mathcal{A}$, for every $n \in \mathbb{N}$, and let $G : S \to ck(\mathbb{R}_0^+)$ be such that the series $\sum_{n=0}^\infty g_i(s_n)\mu_i(\{t_n\}), i = 1, 2$ are convergent. Then G is RL integrable with respect to M and

$$(RL)\int_S G\, dM = \left[\sum_{n=0}^\infty g_1(s_n)\mu_1(\{s_n\}), \sum_{n=0}^\infty g_2(s_n)\mu_2(\{s_n\})\right].$$

Observe moreover that, in this case, the RL-integrability of such G with respect to M implies that the product G · G, as defined in **i**), *is integrable in the same sense. In particular, if such G is a discrete or countable interval-valued signal, the* $(RL)\int_S G \cdot G \, dM$ *represents the energy of the signal.*

If M is of bounded variation and $G : S \to ck(\mathbb{R}_0^+)$ is bounded and such that $G = \{0\}$ M-a.e., then, by [34] (Theorem 3.4), G is M-integrable and $(RL)\int_S G \, dM = \{0\}$.

From now on we suppose that G is bounded and μ_2 is of finite variation.

Proposition 2. *An interval multifunction $G = [g_1, g_2]$ is RL integrable with respect to M on S if and only if g_i are RL integrable with respect to μ_i, $i = 1, 2$ and*

$$\int_S G \, dM = \left[(RL)\int_S g_1 d\mu_1, (RL)\int_S g_2 d\mu_2\right]. \tag{3}$$

Proof. Suppose that $G = [g_1, g_2]$ is RL integrable with respect to $M = [\mu_1, \mu_2]$, that means there exists $[a, b] \in ck(\mathbb{R}_0^+)$ such that for every $\varepsilon > 0$, there exists a countable partition P_ε of S, so that for every tagged partition $P = \{(A_n, t_n)\}_{n \in \mathbb{N}}$ of S with $P \geq P_\varepsilon$, the series $\sigma_{G,M}(P)$ is convergent and

$$d_H([u_{G,M}^{(P)}, v_{G,M}^{(P)}], [a, b]) := \max\{|u_{G,M}^{(P)} - a|, |v_{G,M}^{(P)} - b|\} < \varepsilon.$$

By this inequality it follows that

$$\max\{|\sum_{n=1}^\infty g_1(t_n^*)\mu_1(A_n) - a|, |\sum_{n=1}^\infty g_2(t_n)\mu_2(A_n) - b|\} \leq \varepsilon, \quad \forall n \in \mathbb{N},$$

for every tagged partition $P = \{(A_n, t_n)\}_{n \in \mathbb{N}}$ of S with $P \geq P_\varepsilon$ and then g_i are RL integrable with respect to μ_i, $i = 1, 2$. Formula (3) follows from the convexity of the RL integral.

For the converse, for every $\varepsilon > 0$, let $P_{\varepsilon, g_i}, i = 1, 2$ two countable partitions that verify the definition of RL integrability for g_i, $i = 1, 2$. Let P_ε be a countable partition of S with $P_\varepsilon \geq P_{\varepsilon, g_1} \wedge P_{\varepsilon, g_2}$. Then, for every $P := \{B_n, n \in \mathbb{N}\} \geq P_\varepsilon$ and for every $t_n \in B_n$ it is

$$\left|\sum_{n=0}^{+\infty} g_i(t_n)\mu_i(B_n) - (RL)\int_S g_i d\mu_i\right| < \varepsilon, \quad i = 1, 2.$$

Since g_i, $i = 1, 2$ are selections of G this means that

$$d_H\left([u_{G,M}^{(P)}, v_{G,M}^{(P)}], \left[(RL)\int_S g_1 d\mu_1, (RL)\int_S g_2 d\mu_2\right]\right) \leq \varepsilon$$

and then the assertion follows. □

Remark 6. *By Definition 6 and Proposition 2 we obtain the following definitions for the following cases:*

- *If $M = \{\mu\} : \mathcal{A} \to \mathbb{R}_0^+$ is an arbitrary set function and $G = [g_1, g_2]$ with $g_1(s) \leq g_2(s)$ for every $s \in S$ then*

$$\int_S G \, dM = \left[(RL)\int_S g_1 d\mu, (RL)\int_S g_2 d\mu\right].$$

- *If $M = [\mu_1, \mu_2]$ as in (1) and $G = \{g\} : S \to \mathbb{R}_0^+$ then*

$$\int_S G \, dM = \left[(RL)\int_S g \, d\mu_1, (RL)\int_S g \, d\mu_2\right].$$

Proposition 3. *Let G be an interval valued multifuncion. The RL integrability with respect to M is hereditary on subsets $A \in \mathcal{A}$. Moreover G is RL integrable with respect to M on A if and only if $G\chi_A$ (where χ_A is the characteristic function of the set A) is RL integrable with respect to M on S. In this case, for every $A \in \mathcal{A}$,*

$$(RL)\int_A G\, dM = (RL)\int_S G\chi_A\, dM.$$

Proof. Assume that G is RL integrable in S with respect to M. Let $A \in \mathcal{A}$ and denote by $[a,b]$ the integral of G; then, for every $\varepsilon > 0$, there exists a countable partition P_ε of S, such that, for every finer countable partition $P' := \{A_n\}_{n\in\mathbb{N}}$ and for every $t_n \in A_n$ it is

$$d_H\left(\sigma_{G,M}(P'), [a,b]\right) \leq \varepsilon.$$

Let P_0 be a partition such that $P_0 \geq P_\varepsilon \wedge \{A, T \setminus A\}$, and we denote by $P_A \subset P_0$ the corresponding partition of the set A. Let Π_A be a partition of A finer than P_A, and extend it with a common partition of $S \setminus A$ in such a way the new partition is finer than P_ε.

It is possible to prove that $\sigma_{G,M}(\Pi_A)$ satisfy a Cauchy principle in $ck(\mathbb{R}_0^+)$, and so the first claim follows by the completeness of the space. The equality follows from [34] (Theorem 3.2) and Proposition 2. □

Remark 7. *It is easy to see that, if G is RL integrable with respect to M, for every $\alpha \geq 0$ it is:*

(a)   *αG is RL integrable with respect to M and* $(RL)\int_S \alpha G\, dM = \alpha (RL)\int_S G\, dM.$

(b)   *G is RL integrable with respect to αM and* $(RL)\int_S G\, d(\alpha M) = \alpha (RL)\int_S G\, dM.$

Theorem 1. *If G is an interval valued RL integrable with respect to M multifunction, then $I_G : \mathcal{A} \to ck(\mathbb{R}_0^+)$ defined by*

$$I_G(A) := (RL)\int_A G\, dM$$

is a finitely additive multimeasure.

Proof. By Proposition 3 we have that $I_G(A) \in ck(\mathbb{R}_0^+)$ for every $A \in \mathcal{A}$. In order to prove the additivity we can observe that, for every $A, B \in \mathcal{A}$ with $A \cap B = \emptyset$

$$I_G(A \cup B) = (RL)\int_S G\chi_{A\cup B}\, dM = (RL)\int_S (G\chi_A + G\chi_B)\, dM. \quad (4)$$

If we prove that for every pair of interval valued RL integrable with respect to M multifunctions G_1, G_2 we have that

$$(RL)\int_S (G_1 + G_2)\, dM = (RL)\int_S G_1\, dM + (RL)\int_S G_2\, dM \quad (5)$$

the assertion follows. In order to prove formula (5) let $\varepsilon > 0$ be fixed. Since G_1, G_2 are RL integrable with respect to M, for every $\varepsilon > 0$ there exists a countable partition $P_\varepsilon \in \mathcal{P}$ such that for every $P = \{A_n\}_{n\in\mathbb{N}} \geq P_\varepsilon$ and every $t_n \in A_n$, $n \in \mathbb{N}$, the series $\sigma_{G_i,M}(P)$, $i = 1,2$ are convergent and

$$d_H\left(\sigma_{G_i,M}(P), (RL)\int_S G_i\, dM\right) < \frac{\varepsilon}{2}, \quad i = 1, 2.$$

Then $\sigma_{G_1+G_2,M}(P)$ is convergent and, by [48] (Proposition 1.17),

$$d_H\left(\sigma_{G_1+G_2,M}(P), (RL)\int_S G_1\, dM + (RL)\int_S G_2\, dM\right) < \varepsilon.$$

So $G_1 + G_2$ is RL integrable with respect to M and formula (5) is satisfied.

Now applying formula (5) with $G_1 = G\chi_A$, $G_2 = G\chi_B$ to formula (4) we obtain the additivity of I_G. □

The set-valued integral is monotone relative to the order relation "$\preceq$" and the inclusion one, with respect to the interval-valued integrands.

Proposition 4. *If F, G are two RL integrable with respect to M interval valued multifunctions with $F \preceq G$ then, for every $A \in \mathcal{A}$, $I_F(A) \preceq I_G(A)$.*

Proof. We will prove for $A = S$. Let $F(s) := [f_1(s), f_2(s)]$, $G(s) = [g_1(s), g_2(s)]$. By the integrability of F and G we have, by Proposition 2

$$I_F(S) := (RL) \int_S F \, dM = \left[(RL) \int_S f_1 d\mu_1, (RL) \int_S f_2 d\mu_2 \right],$$

$$I_G(S) := (RL) \int_S G \, dM = \left[(RL) \int_S g_1 d\mu_1, (RL) \int_S g_2 d\mu_2 \right].$$

Since $f_i(s) \leq g_i(s)$ for all $s \in S$ and $i = 1, 2$ by [34] (Theorem 3.10) we have that

$$(RL) \int_S f_1 d\mu_1 \leq (RL) \int_S g_1 d\mu_1, \quad (RL) \int_S f_2 d\mu_2 \leq (RL) \int_S g_2 d\mu_2,$$

and so by the weak interval order, iii), we have that $I_F(S) \preceq I_G(S)$. □

Corollary 1. *If F, G, $F \wedge G$, $F \vee G$ are RL integrable with respect to an interval valued multisubmeasure M then, for every $A \in \mathcal{A}$,*

(a) $\quad (RL) \int_S F \wedge G \, dM \preceq I_F(A) \wedge I_G(A)$;

(b) $\quad I_F(A) \vee I_G(A) \preceq (RL) \int_S F \vee G \, dM.$

Proof. Let $F(s) = [f_1(s), f_2(s)]$, $G(s) = [g_1(s), g_2(s)]$, $h_*(s) = \min\{f_1(s), g_1(s)\}$, $h^*(s) = \min\{f_2(s), g_2(s)\}$. By [34] (Theorem 3.10) $(RL) \int_S h_* d\mu_1 \leq \left\{ (RL) \int_S f_1 d\mu_1, (RL) \int_S g_1 d\mu_1 \right\}$ and an analogous result holds for $(RL) \int_S h^* d\mu_2$. So the result given in 1.a) follows from the definition of $\preceq$ and $\wedge$.

The second statement follows analogously. □

Proposition 5. *Let $F, G : S \to ck(\mathbb{R}_0^+)$ be bounded so that F, G are RL integrable with respect to M. If $F \subseteq G$, then $I_F(A) \subseteq I_G(A)$ for all $A \in \mathcal{A}$.*

Proof. As before we will prove for S. Let $\varepsilon > 0$ be arbitrary. Since F, G are RL integrable with respect to M, there exists a countable partition Π_ε of S so that for every other countable partition $\Pi = \{B_n\}_{n \in \mathbb{N}} \in \mathcal{P}$, with $\Pi \geq \Pi_\varepsilon$ and every choise of points $s_n \in B_n$, $n \in \mathbb{N}$, the series

$$\sum_{n=0}^{\infty} F(s_n) \cdot M(B_n), \quad \sum_{n=0}^{\infty} G(s_n) \cdot M(B_n)$$

are convergent and

$$d_H \left(I_F(S), \sum_{n=0}^{\infty} F(s_n) \cdot M(B_n) \right) < \frac{\varepsilon}{3}; \quad d_H \left(I_G(S), \sum_{n=0}^{\infty} G(s_n) \cdot M(B_n) \right) < \frac{\varepsilon}{3}.$$

Then, by the triangular property of the eccess e,

$$e(I_F(S), I_G(S)) \leq d_H\left(I_F(S), \sum_{n=0}^{\infty} F(s_n) \cdot M(B_n)\right) + e(\sum_{n=0}^{\infty} F(s_n) \cdot M(B_n), \sum_{n=0}^{\infty} G(s_n) \cdot M(B_n)) +$$
$$+ d_H\left(\sum_{n=0}^{\infty} G(s_n) \cdot M(B_n), I_G(S)\right) < \frac{2\varepsilon}{3} + e(\sum_{n=0}^{\infty} F(s_n) \cdot M(B_n), \sum_{n=0}^{\infty} G(s_n) \cdot M(B_n)).$$

Since the series $\sum_{n=0}^{\infty} F(s_n) \cdot M(B_n)$ and $\sum_{n=0}^{\infty} G(s_n) \cdot M(B_n)$ are convergent in $ck(\mathbb{R}_0^+)$, and, by hypothesis, $\sum_{n=0}^{\infty} F(s_n) \cdot M(B_n) \subseteq \sum_{n=0}^{\infty} G(s_n) \cdot M(B_n)$, then

$$e(\sum_{n=0}^{\infty} F(s_n) \cdot M(B_n), \sum_{n=0}^{\infty} G(s_n) \cdot M(B_n)) = 0.$$

Consequently, from the arbitrariety of $\varepsilon > 0$, $e(I_F(S), I_G(S)) = 0$, which implies $I_F(S) \subseteq I_G(S)$. □

We can observe moreover that

Proposition 6. *If G is bounded and RL integrable with respect to M, with M of bounded variation, then*

(a) $\quad \|I_G(S)\| = (RL)\int_S g_2 \, d\mu_2 = (RL)\int_S \|G\| \, d\|M\|.$

(b)
$$\overline{I}_G(S) = \sup\{\sum_{i=1}^{n} |I_G(A_i)|, \{A_i, i=1,\ldots,n\} \in \mathcal{P}\} =$$
$$= \sup\{\sum_{i=1}^{n} (RL)\int_{A_i} g_2 \, d\mu_2, \{A_i, i=1,\ldots,n\} \in \mathcal{P}\} = (RL)\int_S g_2 \, d\mu_2.$$

Proof. It is a consequence of the properties of d_H and [34] (Proposition 3.3, Theorem 3.5). □

Proposition 7. *Let $G : S \to ck(\mathbb{R}_0^+)$ be a bounded multifunction such that G is RL integrable with respect to M on every set $A \in \mathcal{A}$.*

(a) *If M is of bounded variation, then $I_G \ll \overline{M}$ (in the ε-δ sense) and I_G is of finite variation.*
(b) *If moreover M is o-continuous (exhaustive respectively), then I_G is also o-continuous (exhaustive respectively).*

Proof. The statements easily follow by Proposition 6. □

Moreover

Theorem 2. *Let $G : S \to ck(\mathbb{R}_0^+)$ be a multifunction such that G is RL integrable with respect to M on every set $A \in \mathcal{A}$. The following statements hold:*

(a) *If M is monotone, then I_G is monotone too.*
(b) *If M is a d_H-multimeasure of bounded variation then I_G is countably additive.*

Proof. Let $A, B \in \mathcal{A}$ with $A \subseteq B$. By monotonicity $\mu_i(A) \leq \mu_i(B)$ for $i = 1, 2$. We divide B in $A, B \setminus A$ and we apply [34] (Theorem 3.2, Corollary 3.6). The conclusion follows by **(iii)**.

Since M is a d_H-multimeasure, then $\overline{M}$ is countably additive too and o-continuous. Applying Proposition 7 I_G is o-continuous too. Let $(A_n)_{n \in \mathbb{N}} \subset \mathcal{A}$ be an arbitrary sequence of pairwise disjoint sets, with $\bigcup_{n=1}^{\infty} A_n = A \in \mathcal{A}$. We denote by B_n the set $B_n := \bigcup_{k=n+1}^{\infty} A_k$. Since $B_n \searrow \emptyset$, then $\lim_{n\to\infty} \|I_G(B_n)\| = 0$. Since I_G is finitely additive, we have

$$\lim_{n\to\infty} d_H(I_G(A), \sum_{k=1}^{n} I_G(A_k)) = \lim_{n\to\infty} d_H(\sum_{k=1}^{n} I_G(A_k) + I_G(B_n), \sum_{k=1}^{n} I_G(A_k)) \le \lim_{n\to\infty} \|I_G(B_n)\| = 0$$

which ensures that I_G is a d_H-multimeasure. □

Proceeding as in to the proof of the formula (5) and applying [34] (Theorem 3.8) we obtain the following result:

Proposition 8. *Let be $M_1, M_2 : \mathcal{A} \to ck(\mathbb{R}_0^+)$, with $M_1(\emptyset) = M_2(\emptyset) = \{0\}$ and suppose $G : S \to ck(\mathbb{R}_0^+)$ is RL integrable with respect to both M_1 and M_2. If $M : \mathcal{A} \to ck(\mathbb{R}_0^+)$ is the interval-valued multisubmeasure defined by $M(A) = M_1(A) + M_2(A)$, for every $A \in \mathcal{A}$, then G is RL integrable with respect to M and*

$$(RL)\int_S G\, d(M_1 + M_2) = (RL)\int_S G\, dM_1 + (RL)\int_S G\, dM_2.$$

Theorem 3. *Let M be of bounded variation and $F, G : T \to ck(\mathbb{R}_0^+)$ be bounded interval-valued multifunctions. If F, G are RL integrable with respect to M, then*

$$d_H\left((RL)\int_S F\, dM, (RL)\int_S G\, dM\right) \le \sup_{s \in S} d_H(F(s), G(s)) \cdot \overline{M}(S).$$

Proof. Since F, G are M-integrable then f_1, g_1 are μ_1-integrable and f_2, g_2 are μ_2-integrable functions. According to [34] (Theorem 3.9), we have for $i = 1, 2$,

$$\left| (RL)\int_S f_i\, d\mu_i - (RL)\int_S g_i\, d\mu_i \right| \le \sup_{s \in S} |f_i(s) - g_i(s)| \overline{\mu}_i(S). \quad (6)$$

Therefore, by (6) and Remark 5, it follows

$$d_H\left((RL)\int_S F\, dM, (RL)\int_S G\, dM\right) = \max\left\{ \left|(RL)\int_S f_1 d\mu_1 - (RL)\int_S g_1 d\mu_1\right|, \left|(RL)\int_S f_2 d\mu_2 - (RL)\int_S g_2 d\mu_2\right| \right\}$$

$$\le \max\left\{ \sup_{s\in S}|f_1(s) - g_1(s)|\overline{\mu_1}(S), \sup_{s\in S}|f_2(s) - g_2(s)|\overline{\mu_2}(S) \right\} \le$$

$$\le \max\left\{ \sup_{s\in S}|f_1(s) - g_1(s)|, \sup_{s\in S}|f_2(s) - g_2(s)| \right\} \overline{\mu_2}(S) \le$$

$$= \sup_{s\in S} d_H(F(s), G(s))\overline{M}(S).$$

□

Theorem 4. *Let $M_1, M_2 : \mathcal{A} \to ck(\mathbb{R}_0^+)$ and $G : S \to ck(\mathbb{R}_0^+)$ be RL integrable with respect to both M_1 and M_2. Then*

(a) *If $M_1 \preceq M_2$, then $(RL)\int_S G\, dM_1 \preceq (RL)\int_S G\, dM_2$.*

(b) *If $M_1 \subseteq M_2$, then $(RL)\int_S G\, dM_1 \subseteq (RL)\int_S G\, dM_2$.*

Proof. Let $M_1 := [\mu_*, \mu^*]$ and $M_2 := [\nu_*, \nu^*]$. Both the results are consequences of Theorem 2 and [34] (Theorem 3.11). It is enough to observe that if $M_1 \preceq M_2$ then $\mu_* \le \nu_*$ and $\mu^* \le \nu^*$, while if $M_1 \subseteq M_2$ then $\nu_* \le \mu_* \le \mu^* \le \nu^*$. □

As a particular case of Theorem 4 and Corollary 1 we have that for every G which is RL integrable with respect to both positive submeasures μ_1 and μ_2 then

$$(RL)\int_S G\,d(\mu_1 \wedge \mu_2) \preceq (RL)\int_S G\,d\mu_1 \wedge (RL)\int_S G\,d\mu_2.$$

Moreover a convergence result can be obtained using Proposition 1.

Theorem 5. *Let $G_n = [g_1^{(n)}, g_2^{(n)}]$ be a sequence of bounded RL-integrable interval valued multifunction with respect to $M = [\mu_1, \mu_2]$ such that $G_n \preceq G_{n+1}$ for every $n \in \mathbb{N}$. If M is of bounded variation and there exists a function $G = [g_1, g_2]$ such that:*

(a) $d_H(G_n, G) \to 0$ *uniformly;*

(b) $\sup_n \left\| (RL)\int_S G_n dM \right\| < +\infty,$

then G is RL-integrable with respect to M and

$$\lim_{n\to\infty} d_H\left((RL)\int_S G_n\,dM, (RL)\int_S G\,dM\right) = 0.$$

Proof. Since $G_n \preceq G_{n+1}$ we have that $g_i^{(n)} \uparrow$ for $i = 1,2$, this is a consequence of Proposition 4 and Definition 6. By $d_H(G_n, G) \to 0$ uniformly we have that $\max\{|g_i^{(n)} - g_i|, i = 1,2\}$ converges uniformly to zero. We can use now Proposition 1 and we obtain

$$\lim_{n\to\infty} (RL)\int_S g_i^{(n)} d\mu_i = (RL)\int_S g_i d\mu_i, \quad i = 1,2.$$

For every $\varepsilon > 0$ let $k(\varepsilon) \in \mathbb{N}$ be such that

$$d_H(G(t), G_{k(\varepsilon)}(t)) < \varepsilon \ \forall t \in S, \quad \text{and} \quad \left|(RL)\int_S g_i^{(k(\varepsilon))} d\mu_i - (RL)\int_S g_i d\mu_i\right| < \varepsilon, \ i=1,2.$$

So,

$$d_H\left((RL)\int_S G_{k(\varepsilon)}\,dM, \left[(RL)\int_S g_1 d\mu_1, (RL)\int_S g_2 d\mu_2\right]\right) \leq \varepsilon.$$

Let P_ε be the countable partition of S given by $\bigwedge_{i=1,2} P_{\varepsilon,i}$, (the ones that verify Definition 4 for $g_i^{k(\varepsilon)}, i = 1,2$ respectively). Then, for every countable partition $P = \{A_n\}_{n\in\mathbb{N}}$ of S with $P \geq P_\varepsilon$ and for every $t_n \in A_n$ the series $\sigma_{G,M}(P)$ is convergent and

$$d_H\left(\sigma_{G,M}(P), \left[(RL)\int_S g_1 d\mu_1, (RL)\int_S g_2 d\mu_2\right]\right) < d_H\left(\sigma_{G,M}(P), \sigma_{G_{k(\varepsilon)},M}(P)\right) +$$

$$+ d_H\left(\sigma_{G_{k(\varepsilon)},M}(P), (RL)\int_S G_{k(\varepsilon)}\,dM\right) +$$

$$+ d_H\left((RL)\int_S G_{k(\varepsilon)}\,dM, \left[(RL)\int_S g_1 d\mu_1, (RL)\int_S g_2 d\mu_2\right]\right).$$

From previous inequalities and by the arbitrariety of ε the RL-integrability of G follows. □

Remark 8. *Since this research starts from the papers [34,43], this part ends with a comparison between the two types of integral considered: the RL integral with the Gould one given in [43] (Definition 4.7).*

If the interval-valued multifunction F is bounded and μ_2 is of finite variation then, analogously to Proposition 2 it is, by [43] (Proposition 4.9),

$$(G)\int_S F dM = \left[(G)\int_S f_1 d\mu_1, (G)\int_S f_2 d\mu_2\right].$$

So, the two kinds of integral coincide on bounded interval-valued multifunctions with values in $ck(\mathbb{R}_0^+)$ when μ_i, $i = 1, 2$ are complete countably additive measures by [34] (Proposition 4.5) or μ_i, $i = 1, 2$ are monotone, countably -subadditive by [34] (Theorem 4.7).

Without countable additivity the equivalence does not hold; an example can be constructed using [34] (Example 4.6). In the general case only partial results can be obtained on atoms when μ_i, $i = 1, 2$ are monotone, null additive and satisfy property (σ): the proof follows from [34] (Theorem 4.8).

Accordingly with the comparison between Gould and Birkhoff integrals given in [28] we have that Birkhoff, Gould, RL integrals of the bounded single valued functions agree in the countably additive case, see [28] (Theorem 3.10), while in [43] (Remark 5.5) an analogous comparison is given with the Choquet integral.

A comparison between simple Birkhoff and RL integrabilities, introduced in [23,28], in this non additive setting can be obtained using [34] (Theorem 4.2).

Finally we would like to observe that the Rådström's embedding tell us that $(ck(X), d_H, \subseteq)$, when X finite dimensional, is a near vector space with 0 element and order unit B_X. In this case, using [51] (Theorem 5.1), it is a near vector space (see [51] (Definition 2.1) for its definition) that could be embedded, for example, in ℓ_∞ or in $C(\Omega)$ with Ω compact and Hausdorff in such a way the embedding is an isometric isomorphism which takes into account the ordering on the hyperspace.

If we consider instead $(ck(\mathbb{R}_0^+), d_H, \preceq)$, since in general there is no relation between "$\preceq$" and "$\subseteq$" the Rådström embedding provide only the integrability of the interval-valued functions and does not take the weak interval order into account. For this reason we preferred to give the the construction of the RL integral and the proofs, both related to $\preceq$, independently of the Rådström's embedding.

3.1. Applications of Interval Valued Multifunctions

Now, in order to explain what could be the benefits of this approach we give an example of an application of interval valued multifunctions on interval valued multisubmeasure in image processing. In fact a signal can be modeled as an interval-valued multifunction as in [12]. In fact, when the value of the points can not be assigned with precision, it might be preferable to use a measure-based approach.

The advantage of using the notion of interval-valued multifunction in signal analysis is that this formalism allows to include in a unique framework possible uncertainty or the noise on the value of a point.

This situation usually occurs in signal and image processing when images are derived by a measure process, as happens for instance for biomedical images (in CT images, MR images, etc), and in several other applied sciences. In particular, we can apply this representation to a digital image in such a way:

Example 2. *To each pixel (or to a set of pixels) of the image is associated an interval which measures the round-off error which is that committed on the detection on the signal due by the tolerances and by the limits on computational accuracy of the measurements tools ([52]).*

When we consider subsets of pixels we are taking into account the so-called time-jitter error, *i.e., the error that occur in the measure of a given signal when the sampling values can not be matched exactly at the theoretical node but just in a neighborhood of it (see, e.g., [53]).*

In this sense, if $I = (m_{i,j})$ is the matrix associated to a $n \times m$ static, gray-scale image, we can consider the space $\mathcal{S} := (0, n] \times (0, m] \subset \mathbb{R}^2$, and hence the interval-valued multifunction $U_I : \mathcal{S} \to \mathcal{K}_C^+$ corresponding to I, will be given by:

$$U_I(x) := [u_1(x), u_2(x)], \quad x \in \mathcal{S}.$$

The model of a digital image by an interval-valued multifunction as U_I, and obtained by a certain discretization (algorithm) of an analogue image, allows to control the round-off error in the sense that, the true value assumed by original signal at the pixel x belongs to the interval $[u_1(x), u_2(x)]$, in fact providing a lower and an upper bound on the possible oscillations of the sampled image.

For example, in fractal image coding, the functions u_1 and u_2 represent respectively the lower and upper contraction maps of an image, which take into account of the round-off error in the contraction procedure, and can be chosen as follows:

$$u_1(x) := \alpha_1 u(x) + \beta_1(x), \quad u_2(x) := \alpha_2 u(x) + \beta_2(x), \quad x \in \mathcal{S},$$

where α_i, $i = 1, 2$, are suitable integer scaling parameters, $\beta_i : \mathcal{S} \to \mathbb{N}$, $i = 1, 2$, are suitable functions, and $u : \mathcal{S} \to \mathbb{N}$ is the continuous model associated to the starting image I. The functions u_1 and u_2 provide for each pixel the interval containing the true value of the compressed image.

In particular, in the algorithm considered in [15], the functions u_1 and u_2 are piecewise constant, and for a starting image of 225×225 pixel size, they have been defined as follows:

$$U_I(x) = [u_1(x), u_2(x)] = [u(x) - \beta(x), u(x) + \beta(x)], \quad x \in \mathcal{S}, \tag{7}$$

where:

$$u(x) := m_{i,j}, \quad x \in (i-1, i] \times (j-1, j], \quad i, j = 1, \ldots, 225,$$

and

$$\beta(x) := \begin{cases} 0, & x \in (0, 115] \times (0, 115], \\ 40, & x \in (115, 225] \times (115, 225], \\ 20, & \text{otherwise}. \end{cases} \tag{8}$$

As an example we use the interval-valued multifunction (7) to operate with the well-known image of "Baboon" given in Figure 1 (left); the images generated by u_1 and u_2 using the function β defined in (8) are given in Figure 1 (center and right).

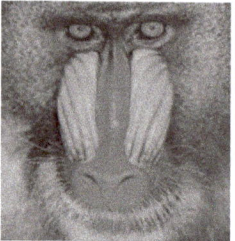  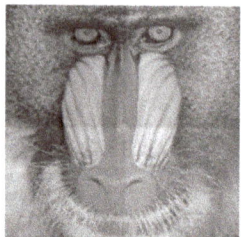

Figure 1. Baboon (**left**); The images generated by u_1 (**center**) and u_2 (**right**) using the interval valued multifunction (7), with β defined in (8).

Here, also numerical truncation have been taken into account, in order to maintain the values of the pixels in the (integer) gray scale $[0, 255]$.

For other examples of functions u_1 and u_2, see, e.g., [13,54]. For instance, in [13] the image representation by multifunctions is used for the implementation of edge detection algorithms, and in this case the corresponding functions u_1 and u_2 are:

$$u_1(x) := \max\left\{0, \min_{x' \in n(x)} \{I(x') - 1\}\right\}, \quad u_2(x) := \min\left\{255, \max_{x' \in n(x)} \{I(x') + 1\}\right\},$$

where $I(x)$ represents the value of a pixel at a position $x \in \mathcal{S}$, while $n(x)$ denotes any set of 3×3 pixels centered at x. For more details, or other applications, see [13,18].

This example was built with the aim to highlight a useful link between the abstract theory of the interval-valued multifunction and the concrete application to image processing. One of the crucial tool in the above set-valued theory is provided by the Hausdorff distance between sets. This special metric plays an important role in the context of digital image processing, where it is used, for example, in order to measure the accuracy of certain class of algorithms, such as those of edge detection, already mentioned in the previous list of possible applications. More precisely, if A is the region of interest (ROI) of a given image and B is the corresponding approximation of the ROI A detected by a suitable edge detection algorithm, the Hausdorff distance measure the displacement between A and B, in fact evaluating the accuracy (i.e., the approximation error) of the method. For instance, in [55] the Hausdorff distance has been used in order to evaluate the degree of accuracy of an algorithm for the detection of the pervious area of the aorta artery from CT images without contrast medium. This procedure is useful, for example, in the diagnosis of aneurysms of the abdominal aorta artery, especially for patients with severe kidneys pathology for which CT images with contrast medium can not be performed. A similar use of the Hausdorff distance could be done for the edge detection algorithms considered in [13,18].

4. Conclusions

A Riemann Lebesgue integral is defined for interval-valued multifunction with respect to interval-valued multisubmeasures. Properties of the integral are established showing in particular that the multimeasure generated is finitely additive. Sufficient conditions for the monotonicity, the order continuity, bounded variation and convergence results are also obtained. A comparison with other integrals is sketchced; an example of an applications in image processing is given highlighting that the advantage of using the notion of interval-valued multifunction in signal analysis is that this formalism allows to include in a unique framework possible uncertainty or the noise on the evaluation of an image at any given pixel. In a future research we will generalize these results in the setting of Banach lattices and we will compare this method with other DIP (digital image processing) algorithms.

Author Contributions: Writing, review and editing, D.C., A.C., A.G., A.I. and A.R.S. All authors have read and agreed to the published version of the manuscript.

Funding: The first and the last authors are members of the working group "Teoria dell'Approssimazione e Applicazioni" of the Italian Mathematical Union (U.M.I.) and they were partially supported by: Grant "Analisi reale, teoria della misura ed approssimazione per la ricostruzione di immagini" (2020) of GNAMPA – INDAM (Italy) and University of Perugia—Fondo Ricerca di Base 2019.

Conflicts of Interest: The authors declare no conflict of interest.

Copyright of the Pictures: The Figure 1 (baboon) are adapted from the original image contained in the repository http://calvados.di.unipi.it/dokuwiki/lib/exe/detail.php?id=ffnamespace:apps&media=ffnamespace:baboon.jpg under the following license: CC Attribution-Share Alike 3.0 Unported.

References

1. Moore, R.E. *Interval Analysis*; Prentice Hall: Englewood Cliffs, NJ, USA, 1966.
2. Weichselberger, K. The theory of interval-probability as a unifying concept for uncertainty. *Int. J. Approx. Reason.* **2000**, *24*, 149–170. [CrossRef]
3. Pap, E. Pseudo-additive measures and their applications. In *Handbook of Measure Theory, II*; Pap, E., Ed.; Elsevier: Amsterdam, The Netherlands, 2002; pp. 1403–1465.
4. Guo, C.; Zhang, D. On set-valued fuzzy measures. *Inform. Sci.* **2004**, *160*, 13–25. [CrossRef]
5. El-Owny, H.; Elaraby, A. Improved Generalized Interval Arithmetic. *Int. J. Math. Comput.* **2013**, *19*, 84–93.
6. Dempster, A.P. Upper and lower probabilities induced by a multivalued mapping. *Ann. Math.Statist.* **1967**, *38*, 325–339. [CrossRef]
7. Zadeh, L.A. Probability measures of fuzzy events. *J. Math. Anal. Appl.* **1968**, *23*, 421–427. [CrossRef]
8. Shafer, G. *A Mathematical Theory of Evidence*; Princeton University Press: Princeton, NY, USA, 1976.
9. Torra, V. Use and Applications of Non-Additive Measures and Integrals. In *Non-Additive Measures, Theory and Applications*; Torra, V., Narukawa, Y., Sugeno, M., Eds.; Studies in Fuzziness and Soft Computing; Springer: Berlin/Heidelberg, Germany, 2014; Volume 310, pp. 1–33. [CrossRef]

10. Torra, V.; Narukawa, Y.; Sugeno, M. (Eds.) *Non-Additive Measures: Theory and Applications*; Studies in Fuzziness and Soft Computing; Springer: Berlin/Heidelberg, Germany, 2014; Volume 310. [CrossRef]
11. Coletti, G.; Petturiti, D.; Vantaggi, B. Models for pessimistic or optimistic decisions under different uncertain scenarios. *Int. J. Approx. Reason.* **2019**, *105*, 305–326. [CrossRef]
12. La Torre, D.; Mendivil, F. Minkowski-additive multimeasures, monotonicity and self-similarity. *Image Anal. Stereol.* **2011**, *30*, 135–142. [CrossRef]
13. Lopez-Molina, C.; De Baets, B.; Barrenechea, E.; Bustince, H. Edge detection on interval-valued images. In *Eurofuse 2011. Advances in Intelligent and Soft Computing*; Melo-Pinto, P., Couto, P., Serôdio, C., Fodor, J., De Baets, B., Eds.; Springer: Berlin/Heidelberg, Germany, 2011; Volume 107. [CrossRef]
14. Vrscay, E.R. A generalized class of fractal-wavelet transforms for image representation and compression. *Can. J. Elect. Comp. Eng.* **1998**, *23*, 69–84. [CrossRef]
15. La Torre, D.; Mendivil, F.; Vrscay, E.R. Iterated function systems on multifunctions. In *Math. Everywhere*; Springer: Berlin/Heidelberg, Germany, 2007; pp. 125–138.
16. Wohlberg, B.; De Jager, G. A review of the fractal image coding literature. *IEEE Trans. Image Process.* **1999**, *8*, 1716–1729. [CrossRef]
17. Zhou, Y.; Zhang, C.; Zhang, Z. An efficient fractal image coding algorithm using unified feature and DCT. *Chaos Solitons Fractals* **2009**, *39*, 1823–1830. [CrossRef]
18. Jurio, A.; Paternain, D.; Lopez-Molina, C.; Bustince, H.; RMesiar, R.; Beliakov, G. A Construction Method of Interval-Valued Fuzzy Sets for Image Processing. In Proceedings of the 2011 IEEE Symposium on Advances in Type-2 Fuzzy Logic Systems (T2FUZZ), Paris, France, 11–15 April 2011.
19. Birkhoff, G. Integration of functions with values in a Banach space. *Trans. Amer. Math. Soc.* **1935**, *38*, 357–378. [CrossRef]
20. Cascales, B.; Rodríguez, J. Birkhoff integral for multi-valued functions. *J. Math. Anal. Appl.* **2004**, *297*, 540–560. [CrossRef]
21. Cascales, B.; Rodríguez, J. The Birkhoff integral and the property of Bourgain. *Math. Ann.* **2005**, *331*, 259–279. [CrossRef]
22. Marraffa, V. A Birkhoff Type Integral and the Bourgain Property in a Locally Convex Space. *Real Anal. Exch.* **2007**, *32*, 409–428. [CrossRef]
23. Potyrala, M. Some remarks about Birkhoff and Riemann-Lebesgue integrability of vector valued functions. *Tatra Mt. Math. Publ.* **2007**, *35*, 97–106.
24. Boccuto, A.; Candeloro, D. Integral and ideals in Riesz spaces. *Inf. Sci.* **2009**, *179*, 2891–2902. [CrossRef]
25. Candeloro, D.; Sambucini, A.R. Order-type Henstock and McShane integrals in Banach lattices setting. In Proceedings of the 2014 IEEE 12th International Symposium on Intelligent Systems and Informatics (SISY), Subotica, Serbia, 11–13 September 2014; pp. 55–59. [CrossRef]
26. Candeloro, D.; Sambucini, A.R. Comparison between some norm and order gauge integrals in banach lattices. *Panam. Math. J.* **2015**, *25*, 1–16.
27. Croitoru, A.; Gavriluţ, A. Comparison between Birkhoff integral and Gould integral. *Mediterr. J. Math.* **2015**, *12*, 329–347. [CrossRef]
28. Candeloro, D.; Croitoru, A.; Gavrilut, A.; Sambucini, A.R. An extension of the Birkhoff integrability for multifunctions. *Mediterranean J. Math.* **2016**, *13*, 2551–2575. [CrossRef]
29. Candeloro, D.; Di Piazza, L.; Musiał, K.; Sambucini, A.R. Gauge integrals and selections of weakly compact valued multifunctions. *J. Math. Anal. Appl.* **2016**, *441*, 293–308. [CrossRef]
30. Caponetti, D.; Marraffa, V.; Naralenkov, K. On the integration of Riemann-measurable vector-valued functions. *Monatsh. Math.* **2017**, *182*, 513–536. [CrossRef]
31. Candeloro, D.; Di Piazza, L.; Musiał, K.; Sambucini, A.R. Some new results on integration for multifunction. *Ric. Mat.* **2018**, *67*, 361–372. [CrossRef]
32. Candeloro, D.; Sambucini, A.R. A Girsanov result through Birkhoff integral. In *International Conference on Computational Science and Its Applications*; Springer International Publishing AG: Cham, Switzerland, 2018, Volume LNCS 10960, pp. 676–683. [CrossRef]
33. Gal, S.G. On a Choquet-Stieltjes type integral on intervals. *Math. Slov.* **2019**, *69*, 801–814. [CrossRef]
34. Candeloro, D.; Croitoru, A.; Gavrilut, A.; Iosif, A.; Sambucini, A.R. Properties of the Riemann-Lebesgue integrability in the non-additive case. *Rend. Circ. Mat. Palermo Ser. 2* **2020**, *69*, 577–589. [CrossRef]

35. Candeloro, D.; Di Piazza, L.; Musiał, K.; Sambucini, A.R. Multi-integrals of finite variation. *Boll. dell'Unione Mat. Ital.* **2020**, *13*, 459–468. [CrossRef]
36. Croitoru, A.; Gavriluţ, A. Convergence results in Birkhoff weak integrability. *Boll. Dell'Unione Mat. Ital.* **2020**, *13*, 477–485. [CrossRef]
37. Di Piazza, L.; Musiał, K. Decompositions of Weakly Compact Valued Integrable Multifunctions. *Mathematics* **2020**, *8*, 863. [CrossRef]
38. Kadets, V.M.; Tseytlin, L.M. On integration of non-integrable vector-valued functions. *Mat. Fiz. Anal. Geom.* **2000**, *7*, 49–65.
39. Kadets, V.M.; Shumyatskiy, B.; Shvidkoy, R.; Tseytlin, L.M.; Zheltukhin, K. Some remarks on vector-valued integration. *Mat. Fiz. Anal. Geom.* **2002**, *9*, 48–65.
40. Gavriluţ, A. Remarks of monotone interval valued set multifunctions. *Inf. Sci.* **2014**, *259*, 225–230. [CrossRef]
41. Iosif, A.; Gavriluţ, A. Integrability in interval-valued (set) multifunctions setting. *Bul. Inst. Politehnic din Iaşi* **2017**, *63*, 65–79.
42. Candeloro, D.; Mesiar, R.; Sambucini, A.R. A special class of fuzzy measures: Choquet integral and applications. *Fuzzy Sets Syst.* **2019**, *355*, 83–99. [CrossRef]
43. Pap, E.; Iosif, A.; Gavriluţ, A. Integrability of an Interval-valued Multifunction with respect to an Interval-valued Set Multifunction. *Iran. J. Fuzzy Syst.* **2018**, *15*, 47–63.
44. Sambucini, A.R. The Choquet integral with respect to fuzzy measures and applications. *Math. Slov.* **2017**, *67*, 1427–1450. [CrossRef]
45. Stamate, C.; Croitoru, A. The general Pettis–Sugeno integral of vector multifunctions relative to a vector fuzzy multimeasure. *Fuzzy Sets Syst.* **2017**, *327*, 123–136. [CrossRef]
46. Torra, V.; Narukawa, Y. On network analysis using non-additive integrals: Extending the game-theoretic network centrality. *Soft Comput.* **2019**, *23*, 2321–2329. [CrossRef]
47. Drewnowski, L. Topological rings of sets, continuous set functions, integration, I, II, III. *Bull. Acad. Polon. Sci. Ser. Math. Astron. Phys.* **1972**, *20*, 277–286.
48. Hu, S.; Papageorgiou, N.S. Handbook of Multivalued Analysis I and II. In *Mathematics and Its Applications, 419*; Kluwer Academic Publisher: Dordrecht, The Netherlands, 1997.
49. Román-Flores, H.; Chalco-Cano, Y.; Lodwick, W.A. Some integral inequalities for interval-valued functions, *Comp. Appl. Math.* **2018**, *37*, 1306–1318.
50. Gavriluţ, A.; Petcu A. Some properties of the Gould type integral with respect to a submeasure. *Bul. Inst. Polit. Iasi Sec. Mat. Mec. Teor. Fiz.* **2007**, *53*, 121–130.
51. Labuschagne, C.C.A.; Pinchuck, A.L.; van Alten, C.J. A vector lattice version of Rådström's embedding theorem. *Quaest. Math.* **2007**, *30*, 285–308. [CrossRef]
52. Butzer, P.L.; Stens, R.L. Linear Prediction by samples from the past. In *Advanced Topics in Shannon Sampling and Interpolation Theory*; Springer: New York, NY, USA, 1993; pp. 157–183.
53. Balakrishnan, A. On the problem of time jitter in sampling. *IRE Trans. Inf. Theory* **1962**, *8*, 226–236. [CrossRef]
54. Mendivil, F.; Vrscay, E.R. Correspondence between fractal-wavelet transforms and iterated function systems with grey-level maps. In *Fractals in Engineering: From Theory to Industrial Applications*; Levy-Vehel, J., Lutton, E., Tricot, C., Eds.; Springer: London, UK, 1997.
55. Costarelli, D.; Seracini, M.; Vinti, G. A segmentation procedure of the pervious area of the aorta artery from CT images without contrast medium. *Math. Methods Appl. Sci.* **2020**, *43*, 114–133. [CrossRef]

Publisher's Note: MDPI stays neutral with regard to jurisdictional claims in published maps and institutional affiliations.

© 2020 by the authors. Licensee MDPI, Basel, Switzerland. This article is an open access article distributed under the terms and conditions of the Creative Commons Attribution (CC BY) license (http://creativecommons.org/licenses/by/4.0/).

Article

Some New Extensions of Multivalued Contractions in a b-metric Space and Its Applications

Reny George [1,2],* and **Hemanth Kumar Pathak [3]**

1. Department of Mathematics, College of Science and Humanities in Alkharj, Prince Sattam bin Abdulaziz University, Al-Kharj 11942, Saudi Arabia
2. Department of Mathematics and Computer Science, St. Thomas College, Bhilai 490009, India
3. SOS in Mathematics, Pt. Ravishankar Shukla University, Raipur 492010, India; hkpathak05@gmail.com
* Correspondence: r.kunnelchacko@psau.edu.sa or renygeorge02@yahoo.com

Abstract: The H^β-Hausdorff–Pompeiu b-metric for $\beta \in [0,1]$ is introduced as a new variant of the Hausdorff–Pompeiu b-metric H. Various types of multi-valued H^β-contractions are introduced and fixed point theorems are proved for such contractions in a b-metric space. The multi-valued Nadler contraction, Czervik contraction, q-quasi contraction, Hardy Rogers contraction, weak quasi contraction and Ciric contraction existing in literature are all one or the other type of multi-valued H^β-contraction but the converse is not necessarily true. Proper examples are given in support of our claim. As applications of our results, we have proved the existence of a unique multi-valued fractal of an iterated multifunction system defined on a b-metric space and an existence theorem of Filippov type for an integral inclusion problem by introducing a generalized norm on the space of selections of the multifunction.

Keywords: b-metric space; H^β-Hausdorff–Pompeiu b-metric; multi-valued fractal; iterated multifunction system; integral inclusion

MSC: 47H10; 47H20; 54H25; 34A60

1. Introduction

Romanian mathematician D. Pompeiu in [1] initiated the study of distance between two sets and introduced the Pompeiu metric. Hausdorff [2] further studied this concept and thereby introduced the Hausdorff–Pompeiu metric H induced by the metric d of a metric space (X, d), as follows:

For any two subsets A and B of X, the function H given by $H(A, B) = \max\{\sup_{x \in A} d(x, B), \sup_{x \in B} d(x, A)\}$ is a metric for the set of compact subsets of X. Note that

$$H(A,B) = \max\{\beta \sup_{x \in A} d(x,B) + (1-\beta) \sup_{x \in B} d(x,A), \beta \sup_{x \in B} d(x,A) + (1-\beta) \sup_{x \in A} d(x,B)\} \text{ for } \beta = 0 \text{ or } 1. \quad (1)$$

Nadler [3] extending the Banach contraction principle introduced multi-valued contraction principle in a metric space using the Hausdorff–Pompieu metric H. Thereafter many extensions and generalizations of multi-valued contraction appeared (see [4–7]). In 1998, Czerwik [8] introduced the Hausdorff–Pompeiu b-metric H_b as a generalization of Hausdorff–Pompeiu metric H and proved the b-metric space version of Nadler contraction principle. Czervik's result drew attention of many researchers who further obtained many generalized multi-valued contractions, named q-quasi contraction [9], Hardy Rogers contraction [10], weak quasi contraction [11], Ciric contraction [12], etc. and proved the existence theorem for such contraction mappings in a b-metric space. The aim of this work is to introduce new variants of the Hausdorff–Pompeiu b-metric and thereby introduce various types of multi-valued H^β-contraction and prove fixed point theorems for such types of contractions in a b-metric space. It is shown that for any b-metric space (X, d_s) and $\beta \in [0, 1]$, the function given in (1) defines a b-metric for the set of closed and bounded subsets of X. We call this metric H^β-Hausdorff–Pompeiu b-metric induced by the b-metric d_s. Thereafter, using this H^β-Hausdorff–Pompeiu b-metric, we have introduced various types of multi-valued H^β-contraction and proved fixed point theorems for such types of contractions in a b-metric space. The multi-valued Nadler contraction [3], Czerwik contraction [8], q-quasi contraction [9], Hardy Rogers contraction [10], Ciric contraction [12], weak quasi contraction [11] existing in literature are all one or the other type of multi-valued H^β-contraction; however, it is shown with proper examples that the converse is not necessarily true. Finally to demonstrate the applications of our results, we prove the existence of a unique multi-valued fractal of an iterated multifunction system defined on a b-metric space and also an existence theorem of Filippov type for an integral inclusion problem by introducing a generalized norm on the space of selections of the multifunction.

2. Preliminaries

Bakhtin [13] introduced b-metric space as follows:

Definition 1 ([13]). *Let X be a nonempty set and $d_s : X \times X \to [0, \infty)$ satisfies:*

1. $d_s(x, y) = 0$ *if and only if* $x = y$ *for all* $x, y \in X$;
2. $d_s(x, y) = d(y, x)$ *for all* $x, y \in X$;
3. *there exist a real number $s \geq 1$ such that $d(x, y) \leq s[d_s(x, z) + d_s(z, y)]$ for all $x, y, z \in X$.*

Then, d_s is called a b-metric on X and (X, d_s) is called a b-metric space with coefficient s.

Example 1. *Let $X = R$ and $d : X \times X \to [0, \infty)$ be given by $d(x, y) = |x - y|^2$, for all $x, y \in X$. Then (X, d) is a b-metric space with coefficient $s = 2$.*

Definition 2 ([13]). *Let (X, d_s) is a b-metric space with coefficient s.*

(i) *A sequence $\{x_n\}$ in X, converges to $x \in X$, if $\lim_{n \to \infty} d_s(x_n, x) = 0$.*
(ii) *A sequence $\{x_n\}$ in X is a Cauchy sequence if for all $\epsilon > 0$, there exist a positive integer $n(\epsilon)$ such that $d_s(x_n, x_m) < \epsilon$ for all $n, m \geq n(\epsilon)$.*
(iii) *(X, d_s) is complete if every Cauchy sequence in X is convergent.*

For some recent fixed point results of single valued and multi-valued mappings in a b-metric space, see [9,14–18]. Throughout this paper, (X, d_s) will denote a complete b-metric space with coefficient s and $CB^{d_s}(X)$ the collection of all nonempty closed and bounded subsets of X with respect to d_s.

For $A, B \in CB^{d_s}(X)$, define $d_s(x, A) = \inf\{d_s(x, a) : a \in A\}$, $\delta_{d_s}(A, B) = \sup_{a \in A} d_s(a, B)$ and $H_{d_s}(A, B) = \max\{\delta_{d_s}(A, B), \delta_{d_s}(B, A)\}$. Czerwik [8] has shown that H_{d_s} is a b-metric in the set $CB^{d_s}(X)$ and is called the Hausdorff–Pompeiu b-metric induced by d_s.

Motivated by the fact that a b-metric is not necessarily continuous (as $\frac{1}{s^2}d_s(x,y) \leq \underline{\lim}_{n \to \infty} d_s(x_n, y_n) \leq \overline{\lim}_{n \to \infty} d_s(x_n, y_n) \leq s^2 d_s(x, y)$ and $\frac{1}{s}d_s(x, y) \leq \underline{\lim}_{n \to \infty} d_s(x_n, y) \leq \overline{\lim}_{n \to \infty} d_s(x_n, y) \leq s d_s(x, y)$ see [19–21]), Miculescu and Mihail [12] introduced the following concept of $*$-continuity.

Definition 3 ([12]). *The b-metric d_s is called $*$-continuous if for every $A \in CB^{d_s}(X)$, every $x \in X$ and every sequence $\{x_n\}$ of elements from X with $\lim_{n \to \infty} x_n = x$, we have $\lim_{n \to \infty} d_s(x_n, A) = d_s(x, A)$.*

Proposition 1 ([17]). *For any $A \subseteq X$,*

$$a \in \bar{A} \iff d_s(a, A) = 0.$$

Lemma 1 ([12]). *Let $\{x_n\}$ be a sequence in (X, d_s). If there exists $\lambda \in [0,1)$ such that $d_s(x_n, x_{n+1}) \leq \lambda d_s(x_{n-1}, x_n)$ for all $n \in \mathbb{N}$, then $\{x_n\}$ is a Cauchy sequence.*

The following lemma can also be proved using the same technique of proof of the above Lemma.

Lemma 2. *Let $\{x_n\}$ be a sequence in (X, d_s). If there exists $\lambda, \epsilon \in [0,1)$, with $\lambda < \epsilon$ such that $d_s(x_n, x_{n+1}) \leq \lambda d_s(x_{n-1}, x_n) + \epsilon^n$ for all $n \in \mathbb{N}$, then $\{x_n\}$ is a Cauchy sequence.*

Czerwik [8] introduced multi-valued contraction in a b-metric space and proved that every multi-valued contraction mapping in a b-metric space has a fixed point.

Definition 4 ([8]). *A mapping $T: X \to CB^{d_s}(X)$ is a multi-valued contraction if there exists $\alpha \in (0, \frac{1}{s})$, such that $g^i, g^j \in X$ implies $H_{d_s}(Tg^i, Tg^j) \leq \alpha\, d_s(g^i, g^j)$.*

Theorem 1 ([8]). *Every multi-valued contraction mapping defined on (X, d_s) has a fixed point.*

Thereafter using Hausdorff–Pompieu b-metric H_{d_s}, many authors introduced several generalized multi-valued contractions in a b-metric space (see Definitions 5 to 8 below) and proved the existence of fixed points for such generalized multi-valued contraction mappings.

Definition 5 ([9]). *A mapping $T: X \to CB^{d_s}(X)$ is a q-multi-valued quasi contraction if there exists $q \in (0, \frac{1}{s})$, such that $g^i, g^j \in X$ implies*

$$H_{d_s}(Tg^i, Tg^j) \leq q \max\{d_s(g^i, g^j), d_s(g^i, Tg^i), d_s(g^j, Tg^j), d_s(g^i, Tg^j), d_s(g^j, Tg^i)\}.$$

Definition 6 ([12]). *A mapping $T: X \to CB^{d_s}(X)$ is a q-multi-valued Ciric contraction if there exists $q, c, d \in (0,1)$, such that $g^i, g^j \in X$ implies*

$$H_{d_s}(Tg^i, Tg^j) \leq q \max\{d_s(g^i, g^j), c\, d_s(g^i, Tg^i), c\, d_s(g^j, Tg^j), \frac{d}{2}(d_s(g^i, Tg^j) + d_s(g^j, Tg^i))\}.$$

Definition 7 ([10]). *A mapping $T: X \to CB^{d_s}(X)$ is a multi-valued Hardy–Roger's contraction if there exists $a, b, c, e, f \in (0,1)$, $a + b + c + 2(e + f) < 1$, such that $g^i, g^j \in X$ implies $H_{d_s}(Tg^i, Tg^j) \leq a\, d_s(g^i, g^j) + b\, d_s(g^i, Tg^i) + c\, d_s(g^j, Tg^j) + e\, d_s(g^i, Tg^j) + f\, d_s(g^j, Tg^i)$.*

Definition 8 ([11]). *A mapping $T: X \to CB^{d_s}(X)$ is a multi-valued weak quasi contraction if there exists $q \in (0,1)$ and $L \geq 0$ such that $g^i, g^j \in X$ implies $H_{d_s}(Tg^i, Tg^j) \leq q \max\{d_s(g^i, g^j), d_s(g^i, Tg^i), d_s(g^j, Tg^j)\} + L\, d_s(g^i, Tg^j)$.*

3. Main Results

3.1. The H^β Hausdorff–Pompieu b-metric

Definition 9. *For $U, V \in CB^{d_s}(X)$, $\beta \in [0,1]$, we define*

$$R^\beta(U, V) = \beta \delta_{d_s}(U, V) + (1 - \beta)\delta_{d_s}(V, U)$$

and

$$H^\beta(U, V) = \max\{R^\beta(U, V), R^\beta(V, U)\}.$$

Proposition 2. *Let $U, V, W \in CB^{d_s}(X)$, we have*
(i) $H^\beta(U, V) = 0$ if and only if $U = V$.
(ii) $H^\beta(U, V) = H^\beta(V, U)$.
(iii) $H^\beta(U, V) \leq s[H^\beta(U, W) + H^\beta(W, V)]$.

Proof. (i) By definition, $H^\beta(U,V) = 0$ implies $\max\{\beta\delta_{d_s}(U,V) + (1-\beta)\delta_{d_s}(V,U), (1-\beta)\delta_{d_s}(U,V) + \beta\delta_{d_s}(V,U)\} = 0$. This gives $\delta_{d_s}(U,V) = 0$ and $\delta_{d_s}(V,U) = 0$. Now, $\delta_{d_s}(U,V) = 0$ implies $d_s(u,V) = 0$ for all $u \in U$. By Proposition 1, we have $u \in \bar{V} = V$ for all $u \in U$ and so $U \subseteq V$. Similarly, $\delta_{d_s}(V,U) = 0$ will imply $V \subseteq U$ and so $U = V$. The reverse implication is clear from the definition.

(ii) Follows from the definition of $H^\beta(U,V)$.

(iii) Let u,v,w be arbitrary elements of U,V,W, respectively. Then we have

$$d_s(u,V) \leq s[d_s(u,w) + d_s(w,V)].$$

Since w is arbitrary, we get

$$d_s(u,V) \leq s[d_s(u,w) + \delta_{d_s}(W,V)] \leq s[d_s(u,W) + \delta_{d_s}(W,V)].$$

Again, since u is arbitrary, we get

$$\delta_{d_s}(U,V) \leq s[\delta_{d_s}(U,W) + \delta_{d_s}(W,V)].$$

Similarly, we have

$$\delta_{d_s}(V,U) \leq s[\delta_{d_s}(V,W) + \delta_{d_s}(W,U)].$$

Therefore,

$$\begin{aligned} R^\beta(U,V) &= \beta\delta_{d_s}(U,V) + (1-\beta)\delta_{d_s}(V,U) \\ &\leq \beta s[\delta_{d_s}(U,W) + \delta_{d_s}(W,V)] + (1-\beta)s[\delta_{d_s}(V,W) + \delta_{d_s}(W,U)] \\ &= s[\beta\delta_{d_s}(U,W) + (1-\beta)\delta_{d_s}(W,U)] + s[\beta\delta_{d_s}(W,V) + (1-\beta)\delta_{d_s}(V,W)] \\ &= s[R^\beta(U,W) + R^\beta(W,V)].\end{aligned}$$

Similarly

$$R^\beta(V,U) \leq s[R^\beta(V,W) + R^\beta(W,U)].$$

Then, we have

$$\begin{aligned} H^\beta(U,V) &= \max\{R^\beta(U,V), R^\beta(V,U)\} \\ &\leq \max\{s[R^\beta(U,W) + R^\beta(W,V)], s[R^\beta(V,W) + R^\beta(W,U)]\} \\ &\leq \max\{sR^\beta(U,W), sR^\beta(W,U)\} + \max\{sR^\beta(W,V), sR^\beta(V,W)\} \\ &= s[H^\beta(U,W) + H^\beta(W,V)]. \end{aligned}$$

□

Remark 1. *In view of Proposition 2, the function $H^\beta : CB^{d_s}(X) \times CB^{d_s}(X) \to [0,+\infty)$, is a b-metric in $CB^{d_s}(X)$ and we call it the H^β-Hausdorff–Pompeiu b-metric induced by d_s.*

Remark 2. *For $\beta \in [0,1]$ $H^\beta(A,B) \leq H_{d_s}(A,B)$ and for $\beta = 0 \vee 1$ $H^\beta(A,B) = H_{d_s}(A,B)$.*

Remark 3. *The Hausdorff–Pompeiu b-metric H^β is equivalent to the Hausdorff–Pompeiu b-metric H_{d_s} in the sense that for any two sets A and B, $H^\beta(A,B) \leq H_{d_s}(A,B) \leq 2H^\beta(A,B)$. However, the examples and applications provided in this paper illustrates the advantages of using H^β-Hausdorff–Pompeiu b-metric in fixed point theory and its applications.*

Theorem 2. *For all $u,v \in X$, $U,V \in CB^{d_s}(X)$ and $\beta \in [0,1]$, the following relations holds:*

(1) $d_s(u,v) = H^\beta(\{u\},\{v\})$,

(2) $U \subset \overline{S}(V, r_1), V \subset \overline{S}(U, r_2) \Rightarrow H^\beta(U, V) \leq r$ where $r = \max\{\beta r_1 + (1-\beta)r_2, \beta r_2 + (1-\beta)r_1\}$,

(3) $H^\beta(U, V) < r \Rightarrow \exists r_1, r_2 > 0$ such that $r = \max\{\beta r_1 + (1-\beta)r_2, \beta r_2 + (1-\beta)r_1\}$ and $U \subset S(V, r_1), V \subset S(U, r_2)$.

Proof. (1) This is immediate from the definition of H^β.
(2) Since $U \subset \overline{S}(V, r_1), V \subset \overline{S}(U, r_2)$, we have that

$$\forall u \in U, \exists v_u \in V \text{ satisfying } d_s(u, v_u) \leq r_1$$

and

$$\forall v \in V, \exists u_v \in U \text{ satisfying } d_s(u_v, v) \leq r_2$$

$$\Rightarrow \inf_{v \in V} d_s(u, v) \leq r_1 \text{ for every } u \in U \text{ and } \inf_{u \in U} d_s(u, v) \leq r_2 \text{ for every } v \in V.$$

$$\Rightarrow \sup_{u \in U}\left(\inf_{v \in V} d_s(u, v)\right) \leq r_1 \text{ and } \sup_{v \in V}\left(\inf_{u \in U} d_s(u, v)\right) \leq r_2.$$

Then, $H^\beta(U, V) \leq r$ where $r = \max\{\beta r_1 + (1-\beta)r_2, \beta r_2 + (1-\beta)r_1\}$.
(3) Let $H^\beta(U, V) = k < r$. Then, there is some $k_1, k_2 > 0$ satisfying

$$k = \max\{\beta k_1 + (1-\beta)k_2, \beta k_2 + (1-\beta)k_1\},$$

$$\delta(U, V) = \sup_{u \in U}(\inf_{v \in V} d_s(u, v)) = k_1, \ \delta(V, U) = \sup_{v \in V}(\inf_{u \in U} d_s(u, v)) = k_2.$$

Since $0 < k < r$, we can find $r_1, r_2 > 0$ such that $k_1 < r_1, k_2 < r_2$ and $r = \max\{\beta r_1 + (1-\beta)r_2, \beta r_2 + (1-\beta)r_1\}$. Thus,

$$\inf_{v \in V} d_s(u, v) \leq k_1 < r_1 \text{ for every } u \in U \text{ and } \inf_{u \in U} d_s(u, v)) \leq k_2 < r_2 \text{ for every } v \in V.$$

Then, for any $u \in U$ there is some $v_u \in V$ satisfying

$$d_s(u, v_u) < \inf_{v \in V} d_s(u, v) + r_1 - k_1 \leq r_1.$$

and, for any $v \in V$ there is some $u_v \in U$ satisfying

$$d_s(u_v, v) < \inf_{u \in U} d_s(u, v) + r_2 - k_2 \leq r_2.$$

Thus, for any $u \in U$ and $v \in V$ we have

$$u \in \bigcup_{v \in V} S(v; r_1) \text{ and } v \in \bigcup_{u \in U} S(u; r_2),$$

which implies

$$U \subset S(V, r_1) \text{ and } V \subset S(U, r_2).$$

□

Remark 4. *From Theorem 2 (2) and (3), it follows that the following statements also hold:*
(2′) $U \subset S(V, r_1), V \subset S(U, r_2) \Rightarrow H^\beta(U, V) \leq r$ where $r = \max\{\beta r_1 + (1-\beta)r_2, \beta r_2 + (1-\beta)r_1\}$
and
(3′) $H^\beta(A, B) < r \Rightarrow \exists r_1, r_2 > 0$ such that $r = \max\{\beta r_1 + (1-\beta)r_2, \beta r_2 + (1-\beta)r_1\}$ and $U \subset \overline{S}(V, r_1), V \subset \overline{S}(U, r_2)$.

Theorem 3. *Let $U, V \in CB^{d_s}(X)$ and $\beta \in [0, 1]$. Then the following equalities holds:*
(4) $H^\beta(U, V) = \inf\{r > 0 : U \subset S(V, r_1), V \subset S(U, r_2)\}$;

(5) $H^\beta(U,V) = \inf\{r > 0 : U \subset \bar{S}(V,r_1), U \subset \bar{S}(V,r_2)\}$,
where $r = \max\{\beta r_1 + (1-\beta)r_2, \beta r_2 + (1-\beta)r_1\}$.

Proof. By (2′), we have

$$H^\beta(U,V) \leq \inf\{r > 0 : U \subset S(V,r_1), U \subset S(V,r_2)\}, r = \max\{\beta r_1 + (1-\beta)r_2, \beta r_2 + (1-\beta)r_1\}. \quad (2)$$

Now let $H^\beta(U,V) = k$, and let $t > 0$. Then $H^\beta(U,V) < k+t$. By Condition (3) of Theorem 2 we can find $t_1, t_2 > 0$ with $\max\{\beta t_1 + (1-\beta)t_2, \beta t_2 + (1-\beta)t_1\} = t$ such that $U \subset S(V; k+t_1)$ and $V \subset S(U; k+t_2)$. Thus,

$$\{r > 0 : U \subset S(V,r_1), B \subset S(U,r_2)\} \supset \{k+t : t > 0, U \subset S(V,k+t_1), V \subset S(U,k+t_2)\}.$$

This implies that

$$\inf\{r > 0 : U \subset S(V,r_1), V \subset S(U,r_2)\} \leq \inf\{k+t : t > 0\} = k = H^\beta(U,V).$$

To conclude,

$$H^\beta(U,V) = \inf\{r > 0 : U \subset S(V,r_1), V \subset S(U,r_2)\}, r = \max\{\beta r_1 + (1-\beta)r_2, \beta r_2 + (1-\beta)r_1\}. \quad (3)$$

□

Theorem 4. *If (X, d_s) is a complete b-metric space, then $(CB^{d_s}(X), H^\beta)$ for any $\beta \in [0,1]$ is also complete. Moreover, $C(X)$ is a closed subspace of $(CB^{d_s}(X), H^\beta)$.*

Proof. Suppose (X, d_s) is complete and the sequence $\{A_n\}_{n \in \mathbb{N}}$ in $CB^{d_s}(X)$ is a Cauchy sequence. Let $B = \{x \in X : \forall \epsilon > 0, m \in \mathbb{N}, \exists n \geq m \text{ for which } S(x,\epsilon) \cap A_n \neq \emptyset\}$.

Let $\epsilon > 0$. By definition of Cauchy sequence, we can find $m(\epsilon) \in \mathbb{N}$ for which, $n \geq m(\epsilon)$ implies $H^\beta(A_n, A_{m(\epsilon)}) < \epsilon$. By Theorem 3 (4), $\exists \epsilon_1, \epsilon_2 > 0$ with $\epsilon = \max\{\beta \epsilon_1 + (1-\beta)\epsilon_2, \beta \epsilon_2 + (1-\beta)\epsilon_1\}$ and $m(\epsilon_1), m(\epsilon_2) \in \mathbb{N}$ such that $\min\{m(\epsilon_1), m(\epsilon_2)\} \geq m(\epsilon)$, $A_n \subset S(A_{m(\epsilon_1)}, \epsilon_1)$ for $n \geq m(\epsilon_1)$ and $A_{m(\epsilon_2)} \subset S(A_n, \epsilon_2)$ $n \geq m(\epsilon_2)$. Then we have $B \subset \bar{S}(A_{m(\epsilon_1)}, \epsilon_1)$, and so

(i) $\quad B \subset \bar{S}(A_{m(\epsilon_1)}, 4\epsilon_1)$ holds.

Now set $\bar{\epsilon}_k = \dfrac{\epsilon_1}{2^k}, k \in \mathbb{N}$, and choose $n_k = m(\bar{\epsilon}_k) \in \mathbb{N}$ such that sequence $\{n_k\}_{k \in \mathbb{N}}$ is strictly increasing and

$$H^\beta(A_n, A_{n_k}) < \bar{\epsilon}_k, \forall n \geq n_k.$$

For some $p \in A_{n_0} = A_{m(\epsilon_1)}$, consider the sequence $\{p_{n_k}\}_{k \in \mathbb{N}}$ with $p_{n_0} = p$, $p_{n_k} \in A_{n_k}$ and $d_s(p_{n_k}, p_{n_{k-1}}) < \dfrac{\epsilon_1}{2^{k-2}}$. It follows that the sequence $\{p_{n_k}\}_{k \in \mathbb{N}}$ is a Cauchy sequence in the complete b-metric space (X, d_s) and so converges to some point $l \in X$.

Additionally, $d_s(p_{n_k}, p_{n_0}) < 4\epsilon_1$ implies $d_s(l, p) \leq 4\epsilon_1$ and so $\inf\limits_{y \in B} d_s(p, y) \leq 4\epsilon_1$, that is, $p \in \bar{S}(B, 4\epsilon_1)$, from which we get

(ii) $\quad A_{n_0} \subset \bar{S}(B, 4\epsilon_1)$.

Now, relations (i), (ii) from above and Theorem 2 (2) yields $H^\beta(A_{n_0}, B) \leq 4\epsilon_1$. Since H^β is a b-metric on $CB^{d_s}(X)$, we have

$$H^\beta(A_n, B) \leq s[H^\beta(A_n, A_{n_0}) + H^\beta(A_{n_0}, B)] < 5s\epsilon_1,$$

for any $n \geq m(\epsilon_1) = n_0$. Hence, sequence $\{A_n\}_{n \in \mathbb{N}}$ is convergent and $(CB^{d_s}(X), H^\beta)$ is complete. □

For the second part, consider the Cauchy sequence $\{A_n\}_{n \in \mathbf{N}}$ in $C(X)$ and consequently in $CB^{d_s}(X)$ and converging to some $A \in CB^{d_s}(X)$. Thus, if $\epsilon > 0$ is chosen, we can find $m(\epsilon) \in \mathbf{N}$ for which

$$H^\beta(A_n, A) < \frac{\epsilon}{2} \ \forall n \geq m(\epsilon), n \in \mathbf{N}.$$

Using (4) of Theorem 3, we get $\exists \epsilon_1, \epsilon_2 > 0$ with $\epsilon = \max\{\beta\epsilon_1 + (1-\beta)\epsilon_2, \beta\epsilon_2 + (1-\beta)\epsilon_1\}$ and $m(\epsilon_1), m(\epsilon_2) \in \mathbf{N}$ such that $\min\{m(\epsilon_1), m(\epsilon_2)\} \geq m(\epsilon)$, $A_n \subset S(A, \frac{\epsilon_1}{2})$ for $n \geq m(\epsilon_1)$ and $A \subset S(A_n, \frac{\epsilon_2}{2})$ for $n \geq m(\epsilon_2)$.

For any fixed $n_0 \geq m(\epsilon_2)$, we have, $A \subset S(A_{n_0}, \frac{\epsilon_2}{2})$ and the compactness of A_{n_0} in X (due to which it is also totally bounded) gives us $x_i^{\epsilon_2}, i \in \overline{1, p}$ such that $A_{n_0} \subset \bigcup_{i=1}^{p} S(x_i^{\epsilon_2}, \frac{\epsilon_2}{2})$, whence $A \subset \bigcup_{i=1}^{p} S(x_i^{\epsilon_2}, \epsilon_2)$. Therefore, $A \in C(X)$.

3.2. Applications to Fixed Point Theory

We begin this section by introducing various classes of multi-valued H^β-contractions in a b-metric space:

Definition 10. $T : X \to CB^{d_s}(X)$ is a multi-valued H^β-contraction if we can find $\beta \in [0,1]$ and $k \in (0,1)$, such that

$$H^\beta(Tg^i, Tg^j) \leq k \cdot d_s(g^i, g^j) \text{ for all } g^i, g^j \in X. \tag{4}$$

Definition 11. $T : X \to CB^{d_s}(X)$ is a multi-valued H^β-Ciric contraction if we can find $\beta \in [0,1]$ and $k \in (0, \frac{1}{s})$, such that for all $g^i, g^j \in X$,

$$H^\beta(Tg^i, Tg^j) \leq k \cdot \max\{d_s(g^i, g^j), d_s(g^i, Tg^i), d_s(g^j, Tg^j), \frac{d_s(g^i, Tg^j) + d_s(g^j, Tg^i)}{2s}\}. \tag{5}$$

Definition 12. $T : X \to CB^{d_s}(X)$ is a multi-valued H^β-Hardy–Rogers contraction if we can find $\beta \in [0,1]$ and $a, b, c, e, f \in (0,1)$ with $a + b + s(c + e) + f < 1$, $\min\{s(a+e), s(b+c)\} < 1$ such that for all $g^i, g^j \in X$,

$$H^\beta(Tg^i, Tg^j) \leq a \cdot d_s(g^i, Tg^i) + b \cdot d_s(g^j, Tg^j) + c \cdot d_s(g^i, Tg^j) + e \cdot d_s(g^j, Tg^i) + f \cdot d_s(g^i, g^j). \tag{6}$$

Definition 13. We say that $T : X \to CB^{d_s}(X)$ is a multi-valued H^β-quasi contraction if we can find $\beta \in [0,1]$ and $k \in (0, \frac{1}{s})$, such that for all $g^i, g^j \in X$,

$$H^\beta(Tg^i, Tg^j) \leq k \cdot \max\{d_s(g^i, g^j), d_s(g^i, Tg^i), d_s(g^j, Tg^j), d_s(g^i, Tg^j), d_s(g^j, Tg^i)\}. \tag{7}$$

Definition 14. We say that $T : X \to CB^{d_s}(X)$ is a multi-valued H^β-weak quasi contraction if we can find $\beta \in [0,1]$, $k \in (0, \frac{1}{s})$ and $L \geq 0$, such that for all $g^i, g^j \in X$,

$$H^\beta(Tg^i, Tg^j) \leq k \cdot \max\{d_s(g^i, g^j), d_s(g^i, Tg^i), d_s(g^j, Tg^j)\} + L d_s(g^j, Tg^i). \tag{8}$$

Example 2. Let $X = [0, \frac{7}{9}] \cup \{1\}$ and $d_s(g^i, g^j) = |g^i - g^j|^2$ for all $g^i, g^j \in X$.

Then $\{X, d_s\}$ is a b-metric space. Define the mapping $T : X \to CB^{d_s}(X)$ by

$$T(g^I) = \begin{cases} \{\frac{g^I}{4}\}, & \text{for } g^I \in [0, \frac{7}{9}] \\ \{0, \frac{1}{3}, \frac{5}{12}\}, & \text{for } g^I = 1. \end{cases}$$

Then T is a multi-valued H^β-contraction with $\beta = \frac{3}{4}$ and $\frac{217}{256} \leq k < 1$ as shown below.
We will consider the following different cases for the elements of X.

(i) $g^I, g^J \in [0, \frac{7}{9}]$.

By Theorem 2(1), we have $H^{\frac{3}{4}}(Tg^I, Tg^J) = d_s(\frac{g^I}{4}, \frac{g^J}{4}) \leq k d_s(g^I, g^J)$, $k \geq \frac{1}{16}$.

(ii) $g^I \in [0, \frac{7}{9}], g^J = 1$.

We have the following sub cases:

(ii)(a) $g^I \in [0, \frac{2}{3}], g^J = 1$. Then $Tg^I = \{\frac{g^I}{4}\}$ and $0 \leq \frac{g^I}{4} \leq \frac{1}{6}$. Therefore, we have $\delta_{d_s}(Tg^I, T1) = \delta_{d_s}(\{\frac{g^I}{4}\}, \{0, \frac{1}{3}, \frac{5}{12}\})$ and $\delta_{d_s}(T1, Tg^I) = \delta_{d_s}(\{0, \frac{1}{3}, \frac{5}{12}\}, \{\frac{g^I}{4}\})$. Note that for $0 \leq \frac{g^I}{4} \leq \frac{1}{6}, \frac{g^I}{4}$ is nearest to 0 and farthest from $\frac{5}{12}$. Therefore, $\delta_{d_s}(Tg^I, T1) = |\frac{g^I}{4} - 0|^2 = \frac{g^{I2}}{16}$ and $\delta_{d_s}(T1, Tg^I) = |\frac{5}{12} - \frac{g^I}{4}|^2 = \frac{9g^{I2} - 30g^I + 25}{144}$
Therefore,

$$H^{\frac{3}{4}}(Tg^I, T1) = \max\{\frac{3}{4}\delta_{d_s}(Tg^I, T1) + \frac{1}{4}\delta_{d_s}(T1, Tg^I), \frac{3}{4}\delta_{d_s}(T1, Tg^I) + \frac{1}{4}\delta_{d_s}(Tg^I, T1)\}$$

$$= \max\{\frac{25}{576} - \frac{10g^I}{192} + \frac{4g^{I2}}{64}, \frac{75}{576} - \frac{30g^I}{192} + \frac{4g^{I2}}{64}\}$$

$$= \frac{75}{576} - \frac{30g^I}{192} + \frac{4g^{I2}}{64} \leq k d_s(g^I, 1), k \geq \frac{279}{576}.$$

($\frac{279}{576}$ is the maximum value of k which satisfies the above inequality for different values of g^I in $[0, \frac{2}{3}]$.)

(ii)(b) $g^I \in (\frac{2}{3}, \frac{7}{9}], g^J = 1$.

Then $Tg^I = \{\frac{g^I}{4}\}$ and $\frac{6}{36} < \frac{g^I}{4} \leq \frac{7}{36}$.

Therefore, we have $\delta_{d_s}(Tg^I, T1) = \delta_{d_s}(\{\frac{g^I}{4}\}, \{0, \frac{1}{3}, \frac{5}{12}\})$ and $\delta_{d_s}(T1, Tg^I) = \delta_{d_s}(\{0, \frac{1}{3}, \frac{5}{12}\}, \{\frac{g^I}{4}\})$. Note that for $\frac{6}{36} < \frac{g^I}{4} \leq \frac{7}{36}, \frac{g^I}{4}$ is nearest to $\frac{1}{3}$ and farthest from $\frac{5}{12}$. Therefore, $\delta_{d_s}(Tg^I, T1) = |\frac{g^I}{4} - \frac{1}{3}|^2 = \frac{g^{I2}}{16} - \frac{2g^I}{12} + \frac{1}{9}$ and $\delta_{d_s}(T1, Tg^I) = |\frac{g^I}{4} - \frac{5}{12}|^2 = \frac{g^{I2}}{16} - \frac{10g^I}{48} + \frac{25}{144}$.
Then, we have

$$H^{\frac{3}{4}}(Tg^I, T1) = \max\{\frac{3}{4}\delta_{d_s}(Tg^I, T1) + \frac{1}{4}\delta_{d_s}(T1, Tg^I), \frac{3}{4}\delta_{d_s}(T1, Tg^I) + \frac{1}{4}\delta_{d_s}(Tg^I, T1)\}$$

$$= \max\{\frac{73}{576} - \frac{34g^I}{192} + \frac{4g^{I2}}{64}, \frac{91}{576} - \frac{38g^I}{192} + \frac{4g^{I2}}{64}\}$$

$$= \frac{91}{576} - \frac{38g^I}{192} + \frac{4g^{I2}}{64} \leq k d_s(g^I, 1), k \geq \frac{217}{256}.$$

However, we see that for $g^i = \frac{7}{9}, g^j = 1$,

$$H(T(\frac{7}{9}), T(1)) = \frac{4}{81} = d_s(\frac{7}{9}, 1)$$

and hence T does not satisfy the contraction Condition of Nadler [3] and Czervic [8].

Example 3. *Let* $X = \{0, \frac{1}{4}, 1\}$, $d_s(g^i, g^j) = |g^i - g^j|^2$ *for all* $g^i, g^j \in X$ *and* $T : X \to CB(X)$ *be as follows:* $T(g^i) = \begin{cases} \{0\}, & \text{for } g^i \in \{0, \frac{1}{4}\} \\ \{0, 1\}, & \text{for } g^i = 1, \end{cases}$

We will show that T is a multi-valued H^β-contraction mapping with $\beta \in (\frac{7}{16}, \frac{9}{16})$. *If* $g^i, g^j \in \{0, \frac{1}{4}\}$, *then the result is clear. Suppose* $g^i \in \{0, \frac{1}{4}\}$ *and* $g^j = 1$. *Then* $\delta_{d_s}(Tg^i, T1) = 0$ *and* $\delta_{d_s}(T1, Tg^i) = 1$ *so that* $H^\beta(Tg^i, T1) = \max\{\beta, 1 - \beta\}$. *In addition, we have* $d_s(g^i, 1) = 1$ *or* $\frac{9}{16}$. *If* $\beta \in (\frac{7}{16}, \frac{1}{2}]$, *then* $H^\beta(Tg^i, T1) = 1 - \beta$. *Now* $1 - \beta \in [\frac{8}{16}, \frac{9}{16})$. *Therefore,* $1 - \beta = \frac{16}{9}(1-\beta)\frac{9}{16}$ *and* $1 - \beta < \frac{16}{9}(1-\beta)1$, *that is* $1 - \beta \leq \frac{16}{9}(1-\beta)d_s(g^i, 1)$. *Thus, we have* $H^\beta(Tg^i, T1) = 1 - \beta \leq kd_s(g^i, 1)$, *where* $k = \frac{16}{9}(1-\beta) < 1$. *Similarly if* $\beta \in [\frac{1}{2}, \frac{9}{16})$, *we get* $H^\beta(Tg^i, T1) = \beta \leq kd_s(g^i, 1)$ *where* $k = \frac{16}{9}\beta < 1$. *Thus, T is a multi-valued H^β-contraction.*

However T is not a multi-valued quasi contraction mapping. Indeed, for $g^i = \frac{1}{4}$ *and* $g^j = 1$, *we have*

$$\begin{aligned}H_{d_s}(T(\frac{1}{4}), T(1)) &= \max\{\delta_{d_s}(T(\frac{1}{4}), T1), \delta_{d_s}(T1, T(\frac{1}{4}))\} = 1 \\ &> k \cdot \max\{d_s(\frac{1}{4}, 1), d_s(\frac{1}{4}, T(\frac{1}{4})), d_s(1, T1), d_s(\frac{1}{4}, T1), d_s(1, T(\frac{1}{4}))\}\end{aligned}$$

for any $k \in (0, 1)$. *Therefore, T does not satisfy the contraction conditions given in Definitions 4–7.*

Now we will present our main results in which we establish the existence of fixed points of generalized multi-valued contraction mappings using H^β Hausdorff–Pompeiu b-metric. Hereafter, $\mathcal{F}\{T\}$ will denote the fixed point set of T.

Theorem 5. *Suppose d_s is $*$-continuous and $T : X \to CB^{d_s}(X)$ is a multi-valued mapping satisfying the following conditions:*

(i) There exists $\beta \in [0, 1]$, $a, b, c, e, f, h, j \geq 0$, $a + b + s(c + e + \frac{h}{2}) + f + j < 1$ *and* $\min\{s(a + e + \frac{h}{2}), s(b + c + \frac{h}{2})\} < 1$ *such that for all* $g^i, g^j \in X$,

$$\begin{aligned}H^\beta(Tg^i, Tg^j) &\leq a \cdot d_s(g^i, Tg^i) + b \cdot d_s(g^j, Tg^j) + c \cdot d_s(g^i, Tg^j) + e \cdot d_s(g^j, Tg^i) \\ &+ h \cdot \frac{d_s(g^i, Tg^j) + d_s(g^j, Tg^i)}{2} + j \cdot \frac{d_s(g^i, Tg^i)d_s(g^j, Tg^j)}{1 + d_s(g^i, g^j)} + f \cdot d_s(g^i, g^j).\end{aligned} \quad (9)$$

(ii) For every g^i in X, $g^j \in T(g^i)$ and $\epsilon > 0$, there exists g in $T(g^j)$ satisfying

$$d_s(g^j, g) \leq H^\beta(Tg^i, Tg^j) + \epsilon. \quad (10)$$

Then $\mathcal{F}\{T\} \neq \phi$.

Proof. For some arbitrary $g_0^t \in X$, if $g_0^t \in Tg_0^t$ then $g_0^t \in \mathcal{F}\{T\}$. Suppose $g_0^t \notin Tg_0^t$. Let $g_1^t \in Tg_0^t$. Again, if $g_1^t \in Tg_1^t$ then $g_1^t \in \mathcal{F}\{T\}$. Suppose $g_1^t \notin Tg_1^t$. By (10), we can find $g_2^t \in Tg_1^t$ such that

$$d_s(g_1^t, g_2^t) \leq H^\beta(Tg_0^t, Tg_1^t) + \epsilon.$$

If $g_2^t \in Tg_2^t$ then $g_2^t \in \mathcal{F}\{T\}$. Suppose $g_2^t \notin Tg_2^t$. By (10), we can find $g_3^t \in Tg_2^t$ such that

$$d_s(g_2^t, g_3^t) \leq H^\beta(Tg_1^t, Tg_2^t) + \epsilon^2.$$

In this way we construct the sequence $\{g_n^t\}$ such that $g_n^t \notin Tg_n^t$, $g_{n+1}^t \in Tg_n^t$ and

$$d_s(g_n^t, g_{n+1}^t) \leq H^\beta(Tg_{n-1}^t, Tg_n^t) + \epsilon^n.$$

Then, using (9), we have

$$
\begin{aligned}
d_s(g_n^t, g_{n+1}^t) &\leq H^\beta(Tg_{n-1}^t, Tg_n^t) + \epsilon^n \\
&\leq a \cdot d_s(g_{n-1}^t, Tg_{n-1}^t) + b \cdot d_s(g_n^t, Tg_n^t) + c \cdot d_s(g_{n-1}^t, Tg_n^t) + e \cdot d_s(g_n^t, Tg_{n-1}^t) \\
&\quad + h \cdot \frac{d_s(g_{n-1}^t, Tg_n^t) + d_s(g_n^t, Tg_{n-1}^t)}{2} + j \cdot \frac{d_s(g_{n-1}^t, Tg_{n-1}^t)d_s(g_n^t, Tg_n^t)}{1 + d_s(g_{n-1}^t, g_n^t)} + f \cdot d_s(g_{n-1}^t, g_n^t) + \epsilon^n,
\end{aligned}
$$

that is,

$$(1 - b - sc - j) \cdot d_s(g_n^t, g_{n+1}^t) \leq \left(a + sc + \frac{sh}{2} + f\right) \cdot d_s(g_{n-1}^t, g_n^t) + \epsilon^n. \tag{11}$$

Using symmetry of H^β, we also have

$$(1 - a - se - j) \cdot d_s(g_n^t, g_{n+1}^t) \leq \left(b + se + \frac{sh}{2} + f\right) \cdot d_s(g_{n-1}^t, g_n^t) + \epsilon^n. \tag{12}$$

Adding (11) and (12), we get

$$d_s(g_n^t, g_{n+1}^t) \leq \left(a + b + s(c + e + \frac{h}{2}) + f + j\right) \cdot d_s(g_{n-1}^t, g_n^t) + \epsilon^n.$$

By Lemma 2, the sequence $\{g^t{}_n\}$ is a Cauchy sequence. Completeness of (X, d_s) gives $\lim_{n \to +\infty} d_s(g_n^t, g^{t*}) = 0$ for some $g^{t*} \in X$. We now show that $g^{t*} \in Tg^{t*}$. Suppose, on the contrary, that $g^{t*} \notin Tg^{t*}$. Then,

$$
\begin{aligned}
\beta \cdot \delta_{d_s}(Tg_n^t, Tg^{t*}) &+ (1 - \beta) \cdot \delta_{d_s}(Tg^{t*}, Tg_n^t) \leq H^\beta(Tg_n^t, Tg^{t*}) \\
&\leq a \cdot d_s(g_n^t, Tg_n^t) + b \cdot d_s(g^{t*}, Tg^{t*}) + c \cdot d_s(g_n^t, Tg^{t*}) + e \cdot d_s(g^{t*}, Tg_n^t) \\
&\quad + h \cdot \frac{d_s(g_n^t, Tg^{t*}) + d_s(g^{t*}, Tg_n^t)}{2} + j \cdot \frac{d_s(g_n^t, Tg_n^t)d_s(g^{t*}, Tg^{t*})}{1 + d_s(g_n^t, g^{t*})} + f \cdot d_s(g_n^t, g^{t*}) \\
&\leq a \cdot d_s(g_n^t, g_{n+1}^t) + b \cdot d_s(g^{t*}, Tg^{t*}) + c \cdot d_s(g_n^t, Tg^{t*}) + e \cdot d_s(g^{t*}, g_{n+1}^t) \\
&\quad + h \cdot \frac{d_s(g_n^t, Tg^{t*}) + d_s(g^{t*}, g_{n+1}^t)}{2} + \frac{d_s(g_n^t, g_{n+1}^t)d_s(g^{t*}, Tg^{t*})}{1 + d_s(g_n^t, g^{t*})} + f \cdot d_s(g_n^t, g^{t*}).
\end{aligned}
$$

and using the *-continuity of d_s, we get

$$\liminf_{n \to \infty} \beta \cdot \delta_{d_s}(Tg_n^t, Tg^{t*}) + (1 - \beta) \cdot \delta_{d_s}(Tg^{t*}, Tg_n^t) \leq \left(b + c + \frac{h}{2}\right) \cdot d_s(g^{t*}, Tg^{t*}).$$

Similarly,

$$\liminf_{n \to \infty} \beta \cdot \delta_{d_s}(Tg^{t*}, Tg_n^t) + (1 - \beta) \cdot \delta_{d_s}(Tg_n^t, Tg^{t*}) \leq \left(a + e + \frac{h}{2}\right) \cdot d_s(g^{t*}, Tg^{t*}).$$

It follows that

$$d_s(g^{i*}, Tg^{i*}) = \beta \cdot d_s(g^{i*}, Tg^{i*}) + (1-\beta) \cdot d_s(Tg^{i*}, g^{i*}) \leq s[\beta \cdot \delta_{d_s}(Tg_n^i, Tg^{i*})$$
$$+ (1-\beta) \cdot \delta_{d_s}(Tg^{i*}, Tg_n^i)] + s \cdot d_s(g_{n+1}^i, g^{i*})$$

that is,

$$d_s(g^{i*}, Tg^{i*}) \leq s[\liminf_{n \to \infty} [\beta \, \delta_{d_s}(Tg_n^i, Tg^{i*}) + (1-\beta)\delta_{d_s}(Tg^{i*}, Tg_n^i)]] + s[\liminf_{n \to \infty} d_s(g_{n+1}^i, g^{i*})]$$
$$\leq s(b + c + \frac{h}{2}) d_s(x^*, Tg^{i*})$$

and

$$d_s(Tg^{i*}, g^{i*}) = \beta \cdot d_s(Tg^{i*}, g^{i*}) + (1-\beta) \cdot d_s(g^{i*}, Tg^{i*}) \leq s[\beta \cdot \delta_{d_s}(Tg^{i*}, Tg_n^i)$$
$$+ (1-\beta) \cdot \delta_{d_s}(Tg_n^i, Tg^{i*})] + s \cdot d_s(g^{i*}, g_{n+1}^i)$$

that is,

$$d_s(Tg^{i*}, g^{i*}) \leq s[\liminf_{n \to \infty} [\beta \cdot \delta_{d_s}(Tg^{i*}, Tg_n^i) + (1-\beta) \cdot \delta_{d_s}(Tg_n^i, Tg^{i*})]] + s[\liminf_{n \to \infty} d_s(g^{i*}, g_{n+1}^i)]$$
$$\leq s(a + e + \frac{h}{2}) \cdot d_s(Tg^{i*}, x^*).$$

Since $\min\{s(a+e+\frac{h}{2}), s(c+e+\frac{h}{2})\} < 1$, we get $d_s(g^{i*}, Tg^{i*}) = 0$ which from Proposition 1 implies that $g^{i*} \in \overline{Tg^{i*}}$ and since Tg^{i*} is closed it follows that $g^{i*} \in Tg^{i*}$. □

Remark 5. *Theorem 5 is true even if we replace (9) by any of the following conditions:*
For some $0 \leq k < \frac{1}{s}$,

$$H^\beta(Tg^i, Tg^j) \leq k \cdot \max\{d_s(g^i, g^j), d_s(g^i, Tg^i), d_s(g^j, Tg^j), \frac{d_s(g^i, Tg^j) + d_s(g^j, Tg^i)}{2s},$$
$$\frac{d_s(g^i, Tg^i)d_s(g^j, Tg^j)}{1 + d_s(g^i, g^j)}\}, \tag{13}$$

$$H^\beta(Tg^i, Tg^j) \leq k \cdot \max\{d_s(g^i, g^j), d_s(g^i, Tg^i), d_s(g^j, Tg^j), d_s(g^i, Tg^j),$$
$$d_s(g^j, Tg^i), \frac{d_s(g^i, Tg^i)d_s(g^j, Tg^j)}{1 + d_s(g^i, g^j)}\}\} \tag{14}$$

The following result is a consequence of Theorem 5 and Remark 5:

Corollary 1. *Suppose d_s is *-continuous and $T: X \to CB^{d_s}(X)$ satisfy Condition (10) and any of the following conditions:*
(i) *T is a multi-valued H^β-Ciric contraction.*
(ii) *T is a multi-valued H^β-Hardy–Roger's contraction.*
(iii) *T is a multi-valued H^β-quasi contraction.*
(iv) *T is a multi-valued H^β-weak quasi contraction.*
(v) *T is a multi-valued H^β-contraction.*
Then $\mathcal{F}\{T\} \neq \phi$.

Taking $T : X \to X$ in Corollary 1 (ii) and using Theorem 2 (i), we have the following corollary.

Corollary 2. *Suppose d_s is $*$-continuous and $T : X \to X$. If there exists non-negative real numbers a, b, c, e, f such that $a + b + s(c + e) + f < 1$, $\min\{s(a+e), s(b+c)\} < 1$ and*

$$d_s(Tg^i, T^j) \leq a \cdot d_s(g^i, g^j) + b \cdot d_s(g^i, Tg^i) + c \cdot d_s(g^j, T^j) + e \cdot d_s(g^i, T^j) + f \cdot d_s(g^j, Tg^i), \text{ for all } g^i, g^j \in X, \quad (15)$$

then $\mathcal{F}(T) \neq \phi$.

Remark 6. *For $\beta = 1$, Condition (10) is obviously satisfied and hence, (Theorem 5 [3]), (Theorem 2.1 [8]), (Theorem 2.2 [9]), (Theorem 2.11 [10]), (Theorem 3.1 [12]) and (Theorem 3.1 [11]) are all particular cases of Corollary 1. However, the examples which follow illustrate that the converse is not necessarily true.*

We now furnish the following examples to validate our results.

Example 4. *Let X, d_s and T be as in Example 2. Then, as shown above, T belongs to the class of multi-valued H^β-contraction with $\beta \in (\frac{7}{16}, \frac{9}{16})$ and consequently T satisfies all the contraction conditions given in Definitions 11–14. We will show that T satisfies (10):*

For $g^i \in [0, \frac{7}{9}]$, Tg^i is singleton and so the result is obvious. Now for $g^i = 1$, if $g^j = 0 \in Tg^i$ then $g = 0 \in Tg^j$ will satisfy (10). If $g^j = \frac{1}{3} \in Tg^i$, then $g = \frac{1}{12} \in Tg^j$ and if $g^j = \frac{5}{12} \in Tg^i$ then $g = \frac{5}{48} \in T^j$ will satisfy (10). Thus, T satisfies conditions of Theorem 5 and Corollary 1 and $0, 1 \in \mathcal{F}(T)$.

However, as shown in Example 2, T does not satisfy the contraction condition of Nadler [3] and Czervic [8].

Example 5. *Let X, d_s and T be as in Example 3. Then as shown above, T belongs to the class of multi-valued H^β-contraction with $\beta \in (\frac{7}{16}, \frac{9}{16})$ and consequently T satisfies all the contraction conditions given in Definitions 11–14.*

We will show that T satisfies (10):

For $g^i \in \{0, \frac{1}{4}\}$, Tg^i is singleton and so the result is obvious. Now for $g^i = 1$, if $g^j = 0 \in Tg^i$ then $g = 0 \in Tg^j$ will satisfy (10). If $g^j = 1 \in Tg^i$ then $g = 1 \in Tg^j$ will satisfy (10). Thus, Theorem 5 and Corollary 1 are applicable and $0, 1 \in \mathcal{F}(T)$. However, we see that T does not satisfy the conditions of (Theorem 2.2 [9]), (Theorem 2.11 [10]) and (Theorem 3.1 [12]).

Example 6. *Let $X = \{0, \frac{1}{12}, \frac{1}{3}, \frac{5}{12}, \frac{34}{48}, 1\}$, $d_s(g^i, g^j) = |g^i - g^j|$ for all $g^i, g^j \in X$ and $T : X \to CB^{d_s}(X)$ be as follows:*

$$T(0) = T(\frac{1}{12}) = \{0\}, \quad T(\frac{1}{3}) = T(\frac{5}{12}) = T(\frac{34}{48}) = \{\frac{1}{12}\}, \quad T(1) = \{0, \frac{1}{3}, \frac{34}{48}, 1\}.$$

Then, T is a multi-valued H^β-quasi contraction for $\beta = \frac{3}{4}$ with $\frac{34}{44} \leq k < 1$ as shown below:

(1) If $g^i = \frac{34}{48}$ and $g^j = 1$, then $\delta_{d_s}(T(\frac{34}{48}), T1) = \delta_{d_s}(\{\frac{1}{12}\}, \{0, \frac{1}{3}, \frac{34}{48}, 1\}) = \frac{1}{12}$ and $\delta_{d_s}(T1, T(\frac{34}{48})) = \delta_{d_s}(\{0, \frac{1}{3}, \frac{34}{48}, 1\}, \{\frac{1}{12}\}) = \frac{11}{12}.$

$$H^{\frac{3}{4}}(T(\frac{34}{48}), T1) = \max\{\frac{3}{4}\delta_{d_s}(T(\frac{34}{48}), T1) + \frac{1}{4}\delta_{d_s}(T1, T(\frac{34}{48}), \frac{3}{4}\delta_{d_s}(T1, T(\frac{34}{48})) + \frac{1}{4}\delta_{d_s}(T(\frac{34}{48}), T1)\}$$
$$= \max\{\frac{3}{4}\cdot\frac{1}{12} + \frac{1}{4}\cdot\frac{11}{12}, \frac{3}{4}\cdot\frac{11}{12} + \frac{1}{4}\cdot\frac{1}{12}\} = \frac{34}{48}$$
$$\leq k\frac{44}{48}, \quad \text{for any } k \geq \frac{34}{44}$$
$$= kd_s(1, T(\frac{34}{48}))$$
$$\leq k\max\{d_s(\frac{34}{48}, 1), d_s(\frac{34}{48}, T(\frac{34}{48})), d_s(1, T1), d_s(\frac{34}{48}, T1), d_s(1, T(\frac{34}{48}))\}.$$

(2) If $g^i = \frac{1}{12}$ and $g^j = 1$. $\delta_{d_s}(T(\frac{1}{12}), T1) = \delta_{d_s}(\{0, \{0, \frac{1}{3}, \frac{34}{48}, 1\}\}) = 0$. $\delta_{d_s}(T1, T(\frac{1}{12})) = \delta_{d_s}(\{0, \frac{1}{3}, \frac{34}{48}, 1\}, 0\}) = 1$.

$$H^{\frac{3}{4}}(T(\frac{1}{12}), T1) = \max\{\frac{3}{4}\delta_{d_s}(T(\frac{1}{12}), T1) + \frac{1}{4}\delta_{d_s}(T1, T(\frac{1}{12})), \frac{3}{4}\delta_{d_s}(T1, T(\frac{1}{12})) + \frac{1}{4}\delta_{d_s}(T(\frac{1}{12}), T1)\} = \frac{3}{4}$$
$$\leq k.1, \quad \text{for any } k \geq \frac{3}{4}$$
$$= k \cdot d_s(1, T(\frac{1}{12}))$$
$$\leq k \cdot \max\{d_s(\frac{1}{12}, 1), d_s(\frac{1}{12}, T(\frac{1}{12})), d_s(1, T1), d_s(\frac{1}{12}, T1), d_s(1, T(\frac{1}{12}))\}.$$

(3) If $g^i = \frac{1}{12}$ and $g^j = \frac{1}{3}$, then $\delta_{d_s}(T(\frac{1}{12}), T(\frac{1}{3})) = \delta_{d_s}(\{0, \{\frac{1}{12}\}\}) = \frac{1}{12}$ and $\delta_{d_s}(\frac{1}{3}, T(\frac{1}{12})) = \delta_{d_s}(\{\frac{1}{12}\}, 0\}) = \frac{1}{12}$.

$$H^{\frac{3}{4}}(T(\frac{1}{12}), T(\frac{1}{3})) = \max\{\frac{3}{4}\delta_{d_s}(T(\frac{1}{12}), T(\frac{1}{3})) + \frac{1}{4}\delta_{d_s}(T(\frac{1}{3}), T(\frac{1}{12})), \frac{3}{4}\delta_{d_s}(T(\frac{1}{3}), T(\frac{1}{12})) + \frac{1}{4}\delta_{d_s}(T(\frac{1}{12}), T(\frac{1}{3}))\}$$
$$= \frac{1}{12} \leq k.\frac{4}{12}, \quad \text{for any } k \geq \frac{1}{4}$$
$$= k \cdot d_s(\frac{1}{3}, T(\frac{1}{12}))$$
$$\leq k \cdot \max\{d_s(\frac{1}{12}, \frac{1}{3}), d_s(\frac{1}{12}, T(\frac{1}{12})), d_s(\frac{1}{3}, T(\frac{1}{3})), d_s(\frac{1}{12}, T(\frac{1}{3})), d_s(\frac{1}{3}, T(\frac{1}{12}))\}.$$

For all other values of g^i and g^j, a similar argument as above follows. Thus, T is a multi-valued Π^β quasi-contraction. We will show that T satisfies (10): For $g^i \in \{0, \frac{1}{12}, \frac{1}{3}, \frac{5}{12}, \frac{34}{48}\}$, Tg^i is singleton and so the result is obvious. Now, for $g^i = 1$, if $g^j = 0 \in Tg^i$ then $g = 0 \in Tg^j$ will satisfy (10). If $g^j = \frac{1}{3}$ or $\frac{34}{48} \in Tg^i$ then, $g = \frac{1}{12} \in Tg^j$ will satisfy (10). Thus, Theorem 5 and Corollary 1 are applicable and $0, 1 \in \mathcal{F}(T)$. However, we see that $H(T(\frac{34}{48}), T(1)) = \frac{11}{12}$, where $d(\frac{34}{48}, 1) = \frac{14}{48}$, $d(\frac{34}{48}, T(\frac{34}{48})) = \frac{30}{48}$, $d(1, T(1)) = 0$, $d(\frac{34}{48}, T(1)) = 0$ and $d(1, T(\frac{34}{48}))\} = \frac{11}{12}$ and so T does not satisfy the conditions of (Theorem 2.2 [9]), (Theorem 2.11 [10]), (Theorem 3.1 [12]) and (Theorem 3.1 [11]).

Proposition 3. Let $T_1, T_2 : X \to CB^{d_s}(X)$, satisfy the following:
(3.1) For all $q, r \in \{1, 2\}$, every g^i in X, g^j in $T_q(g^i)$ and $\epsilon > 0$, there exists g in $T_r(g^j)$ satisfying

$$d_s(g^j, g) \leq H^\beta(T_q g^i, T_r g^j) + \epsilon.$$

(3.2) Any of the following conditions holds:
(i) T_1 and T_2 is a multi-valued H^β-Ciric contraction;
(ii) T_1 and T_2 is a multi-valued H^β-quasi contraction;
(iii) T_1 and T_2 is a multi-valued H^β-weak quasi contraction;

Then, for any $u \in \mathcal{F}\{T_q\}$, there exist $w \in \mathcal{F}\{T_r\}$ ($q \neq r$) such that

$$d_s(u,w) \leq \frac{s}{1-k} \sup_{x \in X} H^\beta(T_q x, T_r x),$$

where k is the Lipschitz's constant.

Proof. Let $g_0^t \in \mathcal{F}\{T_1\}$. By (3.1) we can find $g_1^t \in T_2 g_0^t$ such that

$$d_s(g_0^t, g_1^t) \leq H^\beta(T_1 g_0^t, T_2 g_1^t) + \epsilon.$$

By (3.1), choose $g_2^t \in T_2 g_1^t$ such that

$$d_s(g_1^t, g_2^t) \leq H^\beta(T_2 g_0^t, T_2 g_1^t).$$

Inductively, we define sequence $\{g_n^t\}$ such that $g_{n+1}^t \in T_2(g_n^t)$ and

$$d_s(g_n^t, g_{n+1}^t) \leq H^\beta(T_2 g_{n-1}^t, T_2 g_n^t) + \epsilon. \tag{16}$$

Now, following the same technique as in the proof of Theorem 5, we see that the sequence $\{g_n^t\}$ converges to some g_*^t in X and $g_*^t \in \mathcal{F}\{T_2\}$. Since ϵ is arbitrary, taking $\epsilon \to 0$ in (16) we get

$$d_s(g_n^t, g_{n+1}^t) \leq H^\beta(T_2 g_{n-1}^t, T_2 g_n^t).$$

Then, using (Section 3.2), we get

$$d_s(g_n^t, g_{n+1}^t) \leq k^n d_s(g_0^t, g_1^t).$$

Then, we have $d(g_0^t, g_*^t) \leq \sum_{n=0}^{\infty} s^{n+1} d_s(g_{n+1}^t, g_n^t) \leq s(1 + sk + (sk)^2 + \cdots) d_s(g_1^t, g_0^t) \leq \frac{s}{1-sk}(H^\beta(T_2 g_0^t, T_1 g_0^t) + \epsilon)$. Interchanging the roles of T_1 and T_2 and proceeding as above, it gives that for each $g_0^j \in \mathcal{F}\{T_2\}$ there exist $g_1^j \in T_1 g_0^j$ and $g^\ell \in F(T_1)$ such that

$$d(g_0^j, g^\ell) \leq \frac{s}{1-sk}(H^\beta(T_1 g_0^j, T_2 g_0^j) + \epsilon).$$

Now the result follows as $\epsilon > 0$ is arbitrary. □

3.3. Application to Multi-Valued Fractals

Inspiring from some recent works in [18,22,23], we provide an application of our result to multi-valued fractals. Let $P_i : X \to CB^{d_s}(X)$, $i = 1, 2, \cdots n$ be upper semi continuous mappings. Then, $P = (P_1, P_2, \cdots P_n)$ is an iterated multifunction system (in short IMS) defined on the b-metric space (X, d_s). The operator $T_P : CB(X) \to CB(X)$ defined by $T_P(Y) = \bigcup_{i=1}^n P_i(Y)$ is called the extended multifractal operator generated by the IMS $P = (P_1, P_2, \cdots P_n)$. Any non empty compact subset of X which is a fixed point of T_P is called a multi-valued fractal of the iterated multifunction system $P = (P_1, P_2, \cdots P_n)$.

Theorem 6. Let $P_i : X \to CB(X)$, $i = 1, 2, \cdots n$ be upper semi continuous mappings such that for each $i = 1, 2, \cdots n$ the following conditions hold:
We can find $\beta \in [0,1]$ and $a, e \in (0,1)$, $a + 2se < 1$, such that for all $x, y \in X$, $i = 1, 2 \cdots n$

$$H^\beta(P_i x, P_i y) \leq a\, d_s(x,y) + e[d_s(x, P_i y) + d_s(y, P_i x)]. \tag{17}$$

Then,

(i) For all $U_1, U_2 \in CB(X)$, $H^\beta(T_P(U_1), T_P(U_2)) \leq a H^\beta(U_1, U_2) + e[H^\beta(U_1, T_P(U_2)) + H^\beta(U_2, T_P(U_1))]$.

(ii) A unique multi-valued fractal U^* exists for the iterated multifunction system $P = (P_1, P_2, \cdots P_n)$.

Proof. Suppose condition (17) holds. Then, for $U_1, U_2 \in CB(X)$, we have

$$\begin{aligned}
R^\beta(P_i(U_1), P_i(U_2)) &= \beta \delta(P_i(U_1), P_i(U_2)) + (1-\beta)\delta(P_i(U_2), P_i(U_1)) \\
&= \beta \sup_{x \in U_1}(\inf_{y \in U_2} H^\beta(P_i(x), P_i(y))) + \\
&\quad (1-\beta) \sup_{y \in U_2}(\inf_{x \in U_1} H^\beta(P_i(x), P_i(y))) \\
&\leq \beta \sup_{x \in U_1}(\inf_{y \in U_2} \{a\, d_s(x,y) + e[d_s(x, P_i y) + d_s(y, P_i x)]\}) \\
&\quad +(1-\beta) \sup_{y \in U_2}(\inf_{x \in U_1} \{a\, d_s(x,y) + e[d_s(x, P_i y) + d_s(y, P_i x)]\}) \\
&= a H^\beta(U_1, U_2) + e[H^\beta(U_1, P_i(U_2)) + H^\beta(U_2, P_i(U_1))].
\end{aligned}$$

Similarly, we get

$$R^\beta(P_i(U_2), P_i(U_1)) \leq a H^\beta(U_2, U_1) + e[H^\beta(U_2, P_i(U_1)) + H^\beta(U_1, P_i(U_2))].$$

Thus, we have, for $i = 1, 2, \cdots n$,

$$H^\beta(P_i(U_1), P_i(U_2)) \leq a H^\beta(U_1, U_2) + e[H^\beta(U_2, P_i(U_1)) + H^\beta(U_1, P_i(U_2))].$$

Note that

$$H^\beta\left(\bigcup_{i=1}^n P_i(U_1), \bigcup_{i=1}^n P_i(U_2)\right) \leq \max\{H^\beta(P_1(U_1), P_1(U_2)), H^\beta(P_2(U_1), P_2(U_2)), \cdots H^\beta(P_n(U_1), P_n(U_2))\}$$

and so

$$H^\beta(T_P(U_1), T_P(U_2)) \leq a H^\beta(U_1, U_2) + e[H^\beta(U_1, T_P(U_2)) + H^\beta(U_2, T_P(U_1))].$$

Thus, $T_P : CB(X) \to CB(X)$ satisfies the conditions of Corollary 2 in the metric space $\{CB(X), H^\beta\}$, with $b = c = 0$ and $e = f$ and hence has a fixed point U^* in $CB(X)$, which in turn is the unique multi-valued fractal of the iterated multifunction system $P = (P_1, P_2, \cdots P_n)$. □

Remark 7. Since $H^\beta(A, B) \leq H(A, B)$, Theorem 6 is a proper improvement and generalization of (Theorem 3.4 [18]), (Theorem 3.1 [22]) and (Theorem 3.8 [23]).

3.4. Application to Nonconvex Integral Inclusions

We will begin this section by introducing the following generalized norm on a vector space:

Definition 15. Let V be a vector space over the field K. For some $\rho > 0$ and $\gamma \geq 1$, a real valued function $\|.\|_\gamma^\rho : V \to R$ is a generalized (ρ, γ)-norm if for all $x, y \in V$ and $\lambda \in K$

(1) $\|x\|_\gamma^\rho \geq 0$ and $\|x\|_\gamma^\rho = 0$ if and only if $x = 0$.
(2) $\|\lambda x\|_\gamma^\rho \leq |\lambda|^\rho \|x\|_\gamma^\rho$.
(3) $\|x + y\|_\gamma^\rho \leq \gamma [\|x\|_\gamma^\rho + \|y\|_\gamma^\rho]$.

We say that $(V, \|.\|_\gamma^\rho)$ is a generalized (ρ, γ)-normed linear space.

Remark 8. *The following are immediate consequences of the above definition:*
(i) *Every norm is a generalized (ρ, γ)-norm with $\rho = 1$ and $\gamma = 1$.*
(ii) *Every generalized (ρ, γ)-norm induces a b-metric with coefficient γ, given by $d_\gamma(x,y) = \|x - y\|_\gamma^\rho$.*

Example 7. *Every norm defined on a vector space is a generalized (ρ, γ)-norm.*

Example 8. *Let $V = R$. Define $\|x\|_\gamma^\rho = |x|^2$. Then $\|.\|_\gamma^\rho$ is a generalized $(2,2)$-norm.*

Example 9. *Let $V = R^n$. Define $\|x\|_\gamma^\rho = \sum_k |x_k|^p$, $1 \le p < \infty$. Then $\|.\|_\gamma^\rho$ is a generalized $(p, 2^{p-1})$-norm.*

The convergence, Cauchy sequence and completeness in a generalized (ρ, γ)-normed linear space is defined in the same way as that in a normed linear space.

Throughout this section we will use the following notations and functions:
(i) $A = [0, \tau]$, $\tau > 0$.
(ii) $\mathcal{L}(A)$: is the σ-algebra of all Lebesgue measurable subsets of A.
(iii) Z: is a real separable Banach space with the generalized (ρ, γ)-norm $\|.\|_\gamma^\rho$, for some $\rho > 0$ and $\gamma \ge 1$.
(iv) $\mathcal{P}(Z)$: is the family of all nonempty closed subsets of Z.
(v) d_γ is the b-metric induced by the generalized (ρ, γ)-norm $\|.\|_\gamma^\rho$ and H^β is the H^β-Hausdorff–Pompeiu b-metric on $\mathcal{P}(Z)$, induced by the b-metriv d_γ.
(vi) $\mathcal{B}(Z)$: is the collection of all Borel subsets of Z.
(vii) $\mathcal{C}(A, Z)$: is the Banach space of all continuous functions $g(.): A \to Z$ with norm $\|g(.)\|_* = \sup_{t \in A} \|g(t)\|_\gamma^\rho$.
(viii) $\lambda^\ell(.) : A \to Z$.
(ix) $p(.,.) : A \times Z \to Z$.
(x) $Q(.,.) : A \times Z \to \mathcal{P}(Z)$.
(xi) $q(.,.,.) : A \times A \times Z \to Z$.
(xii) $V : \mathcal{C}(A, Z) \to \mathcal{C}(A, Z)$.
(xiii) $\alpha_1, \alpha_2 : A \times A \to (-\infty, +\infty)$.
(xiv) $L_{\lambda^\ell, \sigma}(t) = Q(t, V(x_{\sigma, \lambda^\ell})(t)), x \in Z, \lambda^\ell \in \mathcal{C}(A, Z), \sigma \in \mathcal{L}^1(A, Z)$.
(xv) $S_{\lambda^\ell}(\sigma) = \{\psi(.) \in \mathcal{L}^1(A, Z) : \psi(t) \in L_{\lambda^\ell, \sigma}(t)\}$.
(xvi) $\mathcal{L}^1(A, Z)$: is the Banach space of all integrable functions u: $A \to Z$, endowed with the norm
$$\|u(.)\|_1 = \int_0^T e^{-\alpha(M_4 M_2 + M_5 M_1) M_3 m(t)} \|u(t)\|_\gamma^\rho \, dt,$$
where $m(t) = \int_0^t k(s) \, ds, t \in A$, M_1, M_2, M_3, M_4, M_5 are positive real constants.

It is well known (see [24]) that $L_{\lambda^\ell, \sigma}(t)$ is measurable and $S_\lambda^\ell(\sigma)$ is nonempty with closed values.

We consider the following integral inclusion

$$x^\ell(t) = \lambda^\ell(t) + \int_0^t [\alpha_1(t,s) \, p(t, u(s)) + \alpha_2(t,s) \, q(t, s, u(s))] \, ds \qquad (18)$$

$$u(t) \in Q(t, V(x^\ell)(t)) \quad a.e. \, t \in A. \qquad (19)$$

We will analyze the above problem (18) and (19) under the following assumptions:
(**AS₁**) $Q(\cdot, \cdot)$ is $\mathcal{L}(I) \otimes \mathcal{B}(X)$ measurable.

(**AS$_2$(i)**) There exists $k(\cdot) \in L^1(A, \mathbf{R}_+)$ such that, for almost all $t \in A$, $Q(t, \cdot)$ satisfies

$$H^\beta(Q(t,x), Q(t,y)) \leq k(t)\|x-y\|_\gamma^\rho$$

for all x, y in Z.

(**AS$_2$(ii)**) For all $x, y \in Z$, $\epsilon > 0$, if $w_1 \in Q(t,x)$ then there exists $w_2 \in Q(t,y)$ such that

$$\|w_1(t) - w_2(t)\|_\gamma^\rho \leq H^\beta(Q(t,x), Q(t,y)) + \epsilon.$$

(**AS$_2$(iii)**) For any $\sigma \in \mathcal{L}^1(A, Z)$, $\epsilon > 0$ and $\sigma_1 \in S_{\lambda^\ell}(\sigma)$, there exists $\sigma_2 \in S_{\lambda^\ell}(\sigma_1)$ such that

$$\|\sigma_1 - \sigma_2\|_1 \leq H^\beta(S_{\lambda^\ell}(\sigma), S_{\lambda^\ell}(\sigma_1)) + \epsilon.$$

(**AS$_3$**) The mappings $f: A \times A \times Z \to Z$, $g: A \times Z \to Z$ are continuous, $V: \mathcal{C}(A, Z) \to \mathcal{C}(A, Z)$

and there exist the constants $M_1, M_2, M_3, M_4 > 0$ such that $(AS_3(i))$ and either $(AS_3(ii)(a))$
or $(AS_3(ii)(b))$ holds $\forall t, s \in A, u_1, u_2 \in \mathcal{L}^1(A, Z), x_1, x_2 \in \mathcal{C}(A, Z)$.

(**AS$_3$(i)**) $\|V(x_1)(t) - V(x_2)(t)\|_\gamma^\rho \leq M_3 \|x_1(t) - x_2(t)\|_\gamma^\rho$.

(**AS$_3$(ii)(a)**) $\|q(t,s,u_1(s)) - q(t,s,u_2(s))\|_\gamma^\rho \leq M_1 N(u_1, u_2),$

$$\|p(s, u_1(s)) - p(s, u_2(s))\|_\gamma^\rho \leq M_2 N(u_1, u_2).$$

(**AS$_3$(ii)(b)**) $\|q(t,s,u_1(s)) - q(t,s,u_2(s))\|_\gamma^\rho \leq M_1 n(u_1, u_2),$

$$\|p(s, u_1(s)) - p(s, u_2(s))\|_\gamma^\rho \leq M_2 n(u_1, u_2),$$

where

$$N(u_1, u_2) = \max\{\|u_1(s) - u_2(s)\|_\gamma^\rho, \|u_1(s) - S_{\lambda^\ell}(u_1)\|_\gamma^\rho, \|u_2(s) - S_{\lambda^\ell}(u_2)\|_\gamma^\rho, \|u_1(s) - S_{\lambda^\ell}(u_2)\|_\gamma^\rho, \|u_2(s) - S_{\lambda^\ell}(u_1)\|_\gamma^\rho\},$$

$$n(u_1, u_2) = \max\{\|u_1(s) - u_2(s)\|_\gamma^\rho, \|u_1(s) - S_{\lambda^\ell}(u_1)\|_\gamma^\rho, \|u_2(s) - S_{\lambda^\ell}(u_2)\|_\gamma^\rho\} + K\|u_1(s) - S_{\lambda^\ell}(u_2)\|_\gamma^\rho$$

and

$$\|u(s) - S_\lambda^\ell(v)\|_\gamma^\rho = \inf_{w \in S_{\lambda^\ell}(v)} \|u(s) - w(s)\|_\gamma^\rho.$$

(**AS$_4$**) α_1, α_2 are continuous, $|\alpha_1(t,s)|^\rho \leq M_4$ and $|\alpha_2(t,s)|^\rho \leq M_5$.

Theorem 7. *Suppose assumptions* (AS_1) *to* (AS_4) *hold and let* $\lambda^\ell(\cdot), \mu^\ell(\cdot) \in \mathcal{C}(A, Z)$, $v(\cdot) \in \mathcal{L}^1(A, Z)$ *be such that* $d(v(t), Q(t, V(y^\ell)(t)) \leq l(t)$ *a.e.* $t \in A$, *where* $l(\cdot) \in \mathcal{L}^1(A, \mathbf{R}_+)$ *and* $y^\ell(t) = \mu^\ell(t, u(t)) + \Phi(u)(t)$, $\forall t \in A$ *with* $\Phi(u)(t) = \int_0^t [\alpha_1(t, \tau)p(\tau, u(\tau)) + \alpha_2 q(t, \tau, u(\tau))] d\tau$, $t \in A$. *Then, for every* $\eta > \gamma$ *and* $\epsilon > 0$, *we can find a solution* $x^\ell(\cdot)$ *of the problem* (18) *and* (19) *such that for every* $t \in A$

$$\|x^\ell(t) - y^\ell(t)\| \leq \|\lambda^\ell - \mu^\ell\|_* \left[1 + \frac{\gamma e^{\eta(M_4 M_2 + M_5 M_1) M_3 m(T)}}{\eta - \gamma}\right]$$

$$+ \frac{\gamma \eta}{\eta - \gamma}(M_4 M_2 + M_5 M_1) e^{\eta(M_4 M_2 + M_1) M_3 m(T)} \int_0^T e^{-\eta(M_4 M_2 + M_5 M_1) M_3 m(t)} l(t) dt.$$

Proof. For $\lambda^\ell \in \mathcal{C}(A,Z)$ and $u \in \mathcal{L}^1(A,Z)$, define

$$x^\ell_{u,\lambda^\ell}(t) = \lambda^\ell(t) + \int_0^t [\alpha_1(t,s)\, p(t,u(s)) + \alpha_2(t,s) q(t,s,u(s))]\, ds.$$

Let $\sigma_1, \sigma_2 \in \mathcal{L}^1(A,Z)$, $w_1 \in S_{\lambda^\ell}(\sigma_1)$ and

$$\mathcal{H}(t) := L_{\lambda^\ell,\sigma_2(t)} \cap \left\{ z \in Z : \|w_1(t) - z\| \leq (M_4 M_2 + M_5 M_1) M_3 k(t) \int_0^t N(\sigma_1, \sigma_2)\, ds + \delta \right\}.$$

By assumption ($AS_2(ii)$), we have

$$d_\gamma(w_1(t), L_{\lambda^\ell,\sigma_2}) \leq H^\beta \Big(Q(t, V(x_{\sigma_1,\lambda^\ell})(t)), Q(t, V(x_{\sigma_2,\lambda^\ell})(t)) \Big) + \epsilon$$

$$\leq k(t) \|V(x_{\sigma_1,\lambda^\ell})(t)) - V(x_{\sigma_2,\lambda^\ell})(t))\|^\rho_\gamma + \epsilon$$

$$\leq M_3 k(t) \|x_{\sigma_1,\lambda^\ell}(t) - x_{\sigma_2,\lambda^\ell}(t)\|^\rho_\gamma + \epsilon$$

$$\leq M_3 k(t) \Big[\int_0^t |\alpha_1(t,s)|^\rho \|p(t,\sigma_1(s)) - p(t,\sigma_2(s))\|^\rho_\gamma ds$$

$$+ \int_0^t |\alpha_2(t,s)|^\rho \|q(t,s,\sigma_1(s)) - q(t,s,\sigma_2(s))\|^\rho_\gamma ds \Big] + \epsilon$$

$$\leq M_3 k(t) \Big[(M_4 M_2 + M_5 M_1) \int_0^t N(\sigma_1, \sigma_2) ds \Big] + \epsilon.$$

Since ϵ is arbitrary, we conclude that $\mathcal{H}(\cdot)$ is nonempty, closed, bounded and measurable.

Let $w_2(\cdot)$ be a measurable selector of $\mathcal{H}(\cdot)$. Then, $w_2 \in S_{\lambda^\ell}(\sigma_2)$. If assumption $AS_3(ii)(a)$ is assumed, then we have

$$\|w_1 - w_2\|_1 = \int_0^T e^{-\eta(M_4 M_2 + M_5 M_1) M_3 m(t)} \|w_1(t) - w_2(t)\|^\rho_\gamma dt$$

$$\leq \int_0^T e^{-\eta(M_4 M_2 + M_5 M_1) M_3 m(t)} M_3 k(t) \Big[(M_4 M_2 + M_5 M_1) \int_0^t N(\sigma_1, \sigma_2) ds \Big] dt$$

$$+ \delta \int_0^T e^{-\eta(M_4 M_2 + M_5 M_1) M_3 m(t)} dt$$

$$\leq \frac{1}{\eta} N^1(\sigma_1, \sigma_2) + \delta \int_0^T e^{-\eta(M_4 M_2 + M_5 M_1) M_3 m(t)} dt,$$

where $N^1(\sigma_1, \sigma_2) = \max\{\|\sigma_1 - \sigma_2\|_1, \|\sigma_1 - S_{\lambda^\ell}(\sigma_1)\|_1, \|\sigma_2 - S^\ell_\lambda(\sigma_2)\|_1, \|\sigma_1 - S_{\lambda^\ell}(\sigma_2)\|_1, \|\sigma_2 - S_{\lambda^\ell}(\sigma_1)\|_1\}$. Since δ is arbitrary, we have

$$d_\gamma(w_1, S_{\lambda^\ell}(\sigma_2)) = \inf_{w_2 \in S_{\lambda^\ell}(\sigma_2)} \|w_1 - w_2\|_1 \leq \frac{1}{\eta} N^1(\sigma_1, \sigma_2).$$

Therefore,

$$\delta_\gamma(S_{\lambda^\ell}(\sigma_1), S_{\lambda^\ell}(\sigma_2)) = \sup_{w_1 \in S_{\lambda^\ell}(\sigma_1)} d_\gamma(w_1, S_{\lambda^\ell}(\sigma_2)) \leq \frac{1}{\eta} N^1(\sigma_1, \sigma_2). \tag{20}$$

Similarly, we also get

$$\delta_\gamma(S_{\lambda^\ell}(\sigma_2), S_{\lambda^\ell}(\sigma_1)) = \sup_{w_1 \in S_{\lambda^\ell}(\sigma_1)} d_\gamma(w_1, S_{\lambda^\ell}(\sigma_2)) \leq \frac{1}{\eta} N^1(\sigma_1, \sigma_2). \tag{21}$$

Multiplying (20) by β and (21) by $1 - \beta$ and adding, we get

$$H^\beta(S_{\lambda^\ell}(\sigma_1), S_{\lambda^\ell}(\sigma_2)) \leq \frac{1}{\eta} N^1(\sigma_1, \sigma_2).$$

Thus, $S_{\lambda^\ell}(\cdot)$ is a H^β-quasi contraction on $\mathcal{L}^1(A, Z)$.

Now let
$$\tilde{Q}(t, x) := Q(t, x) + l(t),$$
$$\tilde{M}_{\lambda^\ell, \sigma}(t) := \tilde{Q}(t, V(x_{\sigma, \lambda^\ell})(t)), \quad t \in I,$$
$$\tilde{S}_{\mu^\ell}(\sigma) := \{\psi(\cdot) \in \mathcal{L}^1(A, Z); \psi(t) \in \tilde{L}_{\mu^\ell, \sigma}(t).$$

It is obvious that $\tilde{Q}(\cdot, \cdot)$ satisfies Hypothesis 5.1.

Let $\phi \in S_{\lambda^\ell}(\sigma), \delta > 0$ and define

$$\tilde{\mathcal{H}}(t) := \tilde{L}_{\lambda^\ell, \sigma(t)} \cap \left\{z \in Z : \|\phi(t) - z\| \leq M_3 k(t) \|\lambda^\ell - \mu^\ell\|_* + l(t) + \delta\right\}.$$

Proceeding in the same way as in the case of $\mathcal{H}(\cdot)$ above, we see that $\tilde{\mathcal{H}}(\cdot)$ is measurable, nonempty and has closed values.

Let $\omega(\cdot) \in S_{\mu^\ell}(\sigma)$. Then

$$\|\phi - \omega\|_1 \leq \int_0^T e^{-\eta(M_4 M_2 + M_5 M_1) M_3 m(t)} \|\phi(t) - \omega(t)\|_\gamma^\rho dt$$

$$\leq \int_0^T e^{-\eta(M_4 M_2 + M_5 M_1) M_3 m(t)} [M_3 k(t) \|\lambda^\ell - \mu^\ell\|_* + l(t) + \delta] dt$$

$$= \|\lambda^\ell - \mu^\ell\|_* \int_0^T e^{-\eta(M_4 M_2 + M_5 M_1) M_3 m(t)} M_3 k(t) dt$$

$$+ \int_0^T e^{-\eta(M_4 M_2 + M_5 M_1) M_3 m(t)} l(t) dt + \delta \int_0^T e^{-\eta(M_4 M_2 + M_5 M_1) M_3 m(t)} dt$$

$$\leq \frac{1}{\eta(M_4 M_2 + M_5 M_1)} \|\lambda^\ell - \mu^\ell\|_*$$

$$+ \int_0^T e^{-\eta(M_4 M_2 + M_5 M_1) M_3 m(t)} l(t) dt + \delta \int_0^T e^{-\eta(M_4 M_2 + M_5 M_1) M_3 m(t)} dt.$$

As $\delta \to 0$ we get

$$H^\beta(S_{\lambda^\ell}(\sigma), \tilde{S}_{\mu^\ell}(\sigma)) \leq \frac{1}{\eta(M_4 M_2 + M_5 M_1)} \|\lambda^\ell - \mu^\ell\|_* \quad (22)$$
$$+ \int_0^T e^{-\eta(M_4 M_2 + M_5 M_1) M_3 m(t)} l(t) dt.$$

Since $S_{\lambda^\ell}(\cdot, \cdot)$ and $\tilde{S}_{\mu^\ell}^\ell(\cdot, \cdot)$ are H^β-quasi contractions with Lipschitz constant $\frac{1}{\eta}$ and since $v(\cdot) \in \mathcal{F}\{\tilde{S}_{\mu^\ell}\}$ by Proposition 3 there exists $u(\cdot) \in \mathcal{F}\{S_{\lambda^\ell}\}$ such that

$$\|v - u\|_1 \leq \frac{\gamma \eta}{\eta - \gamma} \sup_{x \in X} H^\beta(\tilde{S}_{\mu^\ell} x, S_{\lambda^\ell} x).$$

Using (22), we have

$$\|v - u\|_1 \leq \frac{\gamma}{(\eta - \gamma)(M_4 M_2 + M_5 M_1)} \|\lambda^\ell - \mu^\ell\|_*$$
$$+ \frac{\gamma \eta}{\eta - \gamma} \int_0^T e^{-\eta(M_4 M_2 + M_5 M_1) M_3 m(t)} l(t) dt. \quad (23)$$

Now let

$$x^\ell(t) = \lambda^\ell(t) + \int_0^t [\alpha_1(t,s)\, p(t,u(s)) + \alpha_2(t,s) q(t,s,u(s))]\, ds.$$

Then, we have

$$\|x^\ell(t) - y^{(t)}\| \leq \|\lambda^\ell(t) - \mu^\ell(t)\| + (M_4 M_2 + M_5 M_1) \int_0^t \|u(s) - v(s)\| ds$$

$$\leq \|\lambda^\ell - \mu^\ell\|_* + (M_4 M_2 + M_5 M_1) e^{\eta(M_4 M_2 + M_5 M_1) M_3 m(T)} \|u - v\|_1.$$

Using (23) we get

$$\|x^\ell(t) - y^\ell(t)\| \leq \|\lambda^\ell - \mu^\ell\|_* \left[1 + \frac{\gamma e^{\eta(M_4 M_2 + M_5 M_1) M_3 m(T)}}{\eta - \gamma}\right]$$

$$+ \frac{\gamma \eta}{\eta - \gamma}(M_4 M_2 + M_5 M_1) e^{\eta(M_4 M_2 + M_1) M_3 m(T)} \int_0^T e^{-\eta(M_4 M_2 + M_5 M_1) M_3 m(t)} I(t) dt.$$

This completes the proof. □

Remark 9. *Since $H^\beta(A,B) \leq H(A,B)$ and the class of generalized (ρ, γ)-norms includes the usual norm $\|.\|$, we note that the hypothesis conditions $AS_2(i)$ and $AS_3(i), (ii)$ are much weaker than the corresponding hypothesis conditions (Hypothesis 2.1 (ii) and (iii)) of [24]).*

3.5. Conclusions

The H^β-Hausdorff–Pompeiu b-metric is introduced as a new tool in metric fixed point theory and new variants of Nadler, Ciric, Hardy–Rogers contraction principles for multi-valued mappings are established in a b-metric space. The examples and applications provided illustrates the advantages of using H^β-Hausdorff–Pompeiu b-metric in fixed point theory and its applications. The new tool of H^β-Hausdorff–Pompeiu b-metric can be utilized by young researchers in extending and generalizing many of the fixed point results for multi-valued mappings existing in literature and investigate how the new tool would enhance, extend and generalize the applications of the fixed-point theory to linear differential and integro-differential equations, nonlinear phenomena, algebraic geometry, game theory, non-zero-sum game theory and the Nash equilibrium in economics.

Author Contributions: Both authors contributed equally in this research. Both authors have read and agreed to the published version of the manuscript.

Funding: This research received no external funding.

Institutional Review Board Statement: Not applicable.

Informed Consent Statement: Not applicable.

Data Availability Statement: Data sharing not applicable.

Acknowledgments: This research is supported by Deanship of Scientific Research, Prince Sattam bin Abdulaziz University, Alkharj, Saudi Arabia. The authors are thankful to the learned reviewers for their valuable suggestions which helped in bringing this paper to its present form.

Conflicts of Interest: The authors declare no conflict of interest.

References

1. Pompeiu, D. Sur la continuite' des fonctions de variables complexes (These). In *Annales de la Faculté des Sciences de Toulouse*; Gauthier-Villars: Paris, France, 1905; Volume 7, pp. 264–315.
2. Hausdorff, F. *Grunclzuege der Mengenlehre*; Viet: Leipzig, Germany, 1914.
3. Nadler, S.B. Multivalued contraction mappings. *Pacific J. Math.* **1969**, *30*, 475–488. [CrossRef]

4. Damjanovic, B.; Samet, B.; Vetro, C. Common fixed point theorems for multi-valued maps. *Acta Math. Sci. Ser. B Engl. Ed.* **2012**, *32*, 818–824. [CrossRef]
5. Kamran, T.; Kiran, Q. Fixed point theorems for multi-valued mappings obtained by altering distances. *Math. Comput. Model.* **2011**, *54*, 2772–2777. [CrossRef]
6. Klim, D.; Wardowski, D. Fixed point theorems for set valued contractions in complete metric spaces. *J. Math. Anal. Appl.* **2007**, *334*, 132–139. [CrossRef]
7. Liu, Z.; Na, X.; Kwun, Y.C.; Kang, S.M. Fixed points of some set valued F-contractions. *J. Nonlinear Sci. Appl.* **2016**, *9*, 579–5805. [CrossRef]
8. Czerwik, S. Nonlinear set-valued contraction mappings in b-metric spaces. *Atti Sem. Mat. Univ. Modena* **1998**, *46*, 263–276.
9. Aydi, H.; Bota, M.F.; Karapinar, E.; Mitrovic, Z. A fixed point theorem for set-valued quasi-contractions in b-metric spaces. *Fixed Point Theory Appl.* **2012**, *2012*, 88. [CrossRef]
10. Mirmostaffae, A.K. Fixed point theorems for set valued mappings in b-metric spaces. *Fixed Point Theory* **2017**, *18*, 305–314. [CrossRef]
11. Hussain, N.; Mitrovic, Z.D. On multi-valued weak quasi-contractions in b-metric spaces. *J. Nonlinear Sci. Appl.* **2017**, *10*, 3815–3823. [CrossRef]
12. Miculescu, R.; Mihail, A. New fixed point theorems for set-valued contractions in b-metric spaces. *J. Fixed Point Theory Appl.* **2015**, *19*, 2153–2163. [CrossRef]
13. Bakhtin, I.A. The contraction mapping principle in quasimetric spaces. *Funct. Anal. Unianowsk Gos. Ped. Inst.* **1989**, *30*, 26–37.
14. Mebawondu, A.A.; Izuchukwu, C.; Aremu, K.O.; Mewomo, O.T. Some fixed point results for a generalized TAC-Suzuki-Berinde type F-contractions in b-metric spaces. *Appl. Math. E-Notes* **2019**, *19*, 629–653.
15. Pathak, H.K.; George, R.; Nabwey, H.A.; El-Paoumy, M.S.; Reshma, K.P. Some generalized fixed point results in a b-metric space and application to matrix equations. *Fixed Point Theory Appl.* **2015**, *2015*, 101. [CrossRef]
16. George, R.; Nabwey, H.A.; Ramaswamy, R.; Radenovic, S. Some Generalized Contraction Classes and Common Fixed Points in b-Metric Space Endowed with a Graph. *Mathematics* **2019**, *7*, 754. [CrossRef]
17. Boriceanu, M.; Petrusel, A.; Rus, I.A. Fixed point theorems for some multi-valued generalized contractions in b-metric spaces. *Int. J. Math. Stat.* **2010**, *6*, 65–76.
18. Chifu, C.; Petrusel, A. Fixed points for multi-valued contractions in b-metric space wit applications to fractals. *Taiwan. J. Math.* **2014**, *18*, 1365–1375. [CrossRef]
19. Mohanta, S.K. Some fixed point theorems using wt-distance in b-metric spaces. *Fasc. Math.* **2015**, *54*, 125–140. [CrossRef]
20. Nashine, H.K.; Kadelburg, Z. Cyclic generalized ϕ-contractions in b-metric spaces and an application to integral equations. *Filomat* **2014**, *28*, 2047–2057. [CrossRef]
21. Roshan, J.R.; Hussain, N.; Sedghi, S; Shobkolaei, N. Suzuki-type fixed point results in b-metric spaces. *Math. Sci.* **2015**, *9*, 153–160. [CrossRef]
22. Boriceanu, M.; Bota, M.; Petrusel, A. Multivalued fractals in b-metric spaces. *Cent. Eur. J. Math.* **2010**, *8*, 367–377. [CrossRef]
23. Chifu, C.; Petrusel, A. Multivalued fractals and generalized multi-valued contractions. *Chaos Solitons Fractals* **2008**, *36*, 203–210. [CrossRef]
24. Cernea, A. Existence for non convex integral inclusions via fixed points. *Archi. Math. (BRNO)* **2003**, *39*, 293–298.

MDPI
St. Alban-Anlage 66
4052 Basel
Switzerland
Tel. +41 61 683 77 34
Fax +41 61 302 89 18
www.mdpi.com

Mathematics Editorial Office
E-mail: mathematics@mdpi.com
www.mdpi.com/journal/mathematics